Reactivity and Structure
Concepts in Organic Chemistry

Volume 6

Editors:

Klaus Hafner Jean-Marie Lehn
Charles W. Rees P. von Ragué Schleyer
Barry M. Trost Rudolf Zahradník

M. L. Bender M. Komiyama

Cyclodextrin Chemistry

With 14 Figures and 37 Tables

Springer-Verlag
Berlin Heidelberg New York 1978

Myron L. Bender
Professor of Chemistry and Biochemistry
Northwestern University
Evanston, IL 60201/USA

Makoto Komiyama
Postdoctoral Research Associate in Chemistry
Northwestern University
Evanston, IL 60201/USA

ISBN-3-540-08577-7 Springer-Verlag Berlin Heidelberg New York
ISBN-0-387-08577-7 Springer-Verlag New York Heidelberg Berlin

Library of Congress Cataloging in Publication Data. Bender, Myron L 1924– Cyclodextrin chemistry. (Reactivity and structure ; v. 6) Includes bibliographical references and indexes. 1. Dextrine. 2. Cyclic compounds. I. Komiyama, Makoto, 1947– joint author. II. Title. III. Series. QD321.B443 547'.782 77-18001

© by Springer-Verlag Berlin Heidelberg 1978
Printed in Germany

Typesetting: Elsner & Behrens, Oftersheim
Printing and binding: Konrad Triltsch, Würzburg
2152/3140-543210

To Dr. Donald A. Speer, who made much of the research behind this book possible, and to our wives, Muriel S. Bender and Mitsuko Komiyama

Preface

Chemistry was at one time completely described in terms of collision theory, in which one molecule collided with another, sometimes producing reaction. Then came the realization that enzymes which are highly efficient catalysts, work by way of prior complexation, often stereospecific, which is then followed by chemical reaction. Thus, systems that exhibit "host-guest" relationships, i.e., that show complexing are being looked at an ever increasing frequency. The cyclodextrins are the first and probably the most important example of compounds that exhibits complex formation. This is a book about the cyclodextrins. There are of course other compounds that exhibit "host-guest" relationships and thus bind other organic molecules, but so far they have not achieved the importance of the cyclodextrins.

By their name it is obvious that cyclodextrins are cyclic compounds. The complexes that they form are therefore cyclic inclusion complexes. Because the complexes are cyclic in nature, complexation can be very strong, as opposed to π-complex, electrostatic, or apolar complexes in which complex formation is two-dimensional rather than three-dimensional.

Cyclodextrins turn out to be excellent models of enzymes. This is probably not fortuitous because they were first sought since it was discovered that the principal binding in the enzyme chymotrypsin was a cyclic inclusion complex.

Cyclodextrins can do more than form cyclic inclusion complexes, they can catalyze as well. But catalysis always occurs after complex formation.

Cyclodextrins can catalyze in a covalent manner after complexation, for then the many hydroxyl groups of the cyclodextrin can act as nucleophiles. Of course it is possible to attach other groups to cyclodextrins which can also act as nucleophiles. In such cases covalent intermediates in which one part of the substrate is attached to the cyclodextrin will be formed. Catalysis can also take the form of non-covalent catalysis in which the inside of these doughnut-shaped molecules acts as a medium of low dielectric constant for reaction to occur or in which the torus of the molecule can preferentially bind one conformer better than another. All these possibilities will be discussed.

We should like to thank the large number of people who have worked on cyclodextrins during the past years. We also thank the National Science Foundation and the Hoffmann-LaRoche Co. who have supported research in this area for many years. Makoto Komiyama would like to thank his former teacher, Dr. Hidefumi Hirai, and his parents, Mr. and Mrs. Kojiro Komiyama. We both thank Ms. Rosalind Bach who did a superb job in preparing this book for publication.

September, 1977

M. L. Bender and M. Komiyama
Northwestern University
Evanston, Illinois, USA.

Contents

Contents

I. Introduction

The cyclodextrins, whose reactions will be discussed in detail here, have been the subject of a book in 1954 [1], eight review articles [2–9] and innumerable papers (the references in this book total over 300). The above-mentioned book put forward for the first time the concept that certain organic molecules can form a cyclic inclusion complex with other molecules. This is what would be called the "host-guest" relationship in modern terminology. This was followed by a review in 1957 [5] that emphasized the preparation, isolation, and separation of these molecules on the basis of size [5]. Subsequently, it was found not only could these molecules form insoluble complexes but further that they could form stoichiometric complexes in dilute aqueous solution. So the way was paved for the use of these compounds as enzyme models [3, 4, 6, 7]. This led to their use as potential catalysts, both in enzymatic and non-enzymatic reactions [2].

Investigations of cyclodextrin chemistry is on the increase in many parts of the world (the U. S. A., Canada, Holland, Germany, and Japan) and it was thought appropriate to write a book on this subject since it is undoubtedly coming of age.

In this book, we will cover (up to March, 1977) the structure of the cyclodextrins, complex formation with cyclodextrins, catalyses by cyclodextrins, and chemical modifications of cyclodextrins to make them better catalysts.

Cyclodextrins are but one example of relatively simple organic compounds which complex other organic molecules. The cyclodextrins are the first, probably the simplest, and certainly the most water soluble. The cyclodextrins are natural products whereas most of the other hosts containing methylene groups [10] or benzenoid systems [11] have had to be synthesized. For this reason alone, the cyclodextrins are more readily available and are used for complexing purposes to a greater extent.

The cyclodextrins have molecular weights around 1000. Thus they are considerably larger than conventional organic compounds. On the other hand, they are considerably smaller than the enzymes which have molecular weights of 10,000 and up (A considerable number of them have molecular weights in the 20,000–30,000 range). Thus the cyclodextrins bridge the gap between ordinary organic compounds and enzymes.

II. Properties

1. Source and Nomenclature

The cyclodextrins, sometimes called Schardinger dextrins, cycloamyloses, or cyclo-
glucans, are a series of oligosaccharides produced by the action of the amylase of
Bacillus macerans on starch and related compounds. Though cyclodextrins were discovered in 1891 by Villiers [12], the first detailed description of their preparation
and isolation was made by Schardinger [13–15].

Cyclodextrins are composed of α-(1,4)-linkages of a number of D(+)-glucopyra-
nose units. Cyclodextrins are designated by a Greek letter to denote the number of
glucose units: α- for 6, β- for 7, γ- for 8 and so on. Sometimes, α-, β-, and γ-cyclodex-
trins are called cyclohexaamylose, cycloheptaamylose, and cyclooctaamylose.

Treatment of starch with the amylase of *Bacillus macerans* (cyclodextrinase)
gives crude starch digests containing α-, β-, and γ-cyclodextrins together with small
amounts of higher cyclodextrins [16–21]. An easy way to separate α-, β-, and γ-cyclo-
dextrins from the digest is selective precipitation by appropriate organic compounds
[17–19]. For example, α-, β-, and γ-cyclodextrins can be precipitated from the digest
by addition of a tetrachloroethylene-tetrachloroethane mixture. Then, addition of
p-cumene to an aqueous solution of three cyclodextrins precipitates only β- and
γ-cyclodextrins, whereas α-cyclodextrin is left in solution together with remaining
β- and γ-cyclodextrins. α-cyclodextrin is isolated by selective precipitation with cyclohexane, β-cyclodextrin with fluorobenzene, and γ-cyclodextrin with anthracene [17].
This method is based on the difference in the sizes of cavities of α-, β-, and γ-cyclo-
dextrins. Absorption chromatography was also used for separation [22]. Furthermore,
recent developments in high temperature cellulose column chromatography enabled
the separation or detection of higher cyclodextrins [20, 21].

Up to now, α-, β- [13], γ- [18], and δ- [18–21] cyclodextrins have been isolated.
In addition to them, ϵ-, ζ-, η-, and θ-cyclodextrins were identified by column chromatography, though they were obtained as mixtures of purely α-(1.4) linked cyclic molecules with small amounts of branched cyclic molecules and branched open-chain dextrins [21]. Cyclodextrins, having fewer than six glucose residues, are not known to
exist, probably because of steric hindrance [23]. A recent patent [24] indicates that
addition of isoamylase increases the yield of cyclodextrins on treatment of starch by
the amylase of *Bacillus macerans.*

2

Methods for the quantitative analysis of mixtures of cyclodextrins are required in connection with the production of cyclodextrins as well as their application, since α-, β-, and γ-cyclodextrins are simultaneously produced from starch together with the higher series of cyclodextrins. High temperature cellulose column chromatography is one of the most effective methods [20, 21]. Gas chromatography is also a direct and accurate method for analysis of a mixture of α- and β-cyclodextrins. However, this method requires conversion of the cyclodextrins to their volatile dimethylsilyl ethers prior to the analysis [25]. ^{1}H—NMR spectroscopy is applicable to analysis of concentrated (10—30%) solution of cyclodextrins in methyl sulfoxide [26]. For analyses of dilute aqueous solutions of mixtures of cyclodextrins, a technique using fluorescent dyes is effective. Cyclodextrins enhances the fluorescence of 2-p-toluidinylnaphthalene-6-sulphonate, the magnitude of which depends largely on the size of cyclodextrins (see p. 5). Thus, the magnitude of the fluorescence enhancement can be related to the concentrations of α- and β-cyclodextrins [27]. Both thin-layer chromatography [28] and circular paper chromatography [16] are also applicable.

2. Structure and Physical Properties

Cyclodextrins have doughnut-shapes with all the glucose units in substantially undistorted C1 (D) (chair) conformations. Figs. 1 and 2 show schematic diagrams as well as Corey-Pauling-Kultun molecular models for cyclodextrins.

These structures require special arrangements of functional groups in cyclodextrin molecules, resulting in a variety of interesting features of cyclodextrins. Thus, the secondary hydroxyl groups (on the C-2 and C-3 atoms of the glucose units) are located on one side of the torus, whereas the primary hydroxyl groups are located on the opposite side of the torus. The interior of the torus consists only of a ring of C—H groups, a ring of glucosidic oxygens and another ring of C—H groups. Therefore, the interior of the toruses of cyclodextrins are relatively apolar compared to water.

Cyclodextrin cavities are slightly "V" shaped with the secondary hydroxyl side more open than the primary hydroxyl side. The primary hydroxyl groups can rotate so as to partially block the cavity, while the secondary hydroxyl groups on relatively rigid chains cannot.

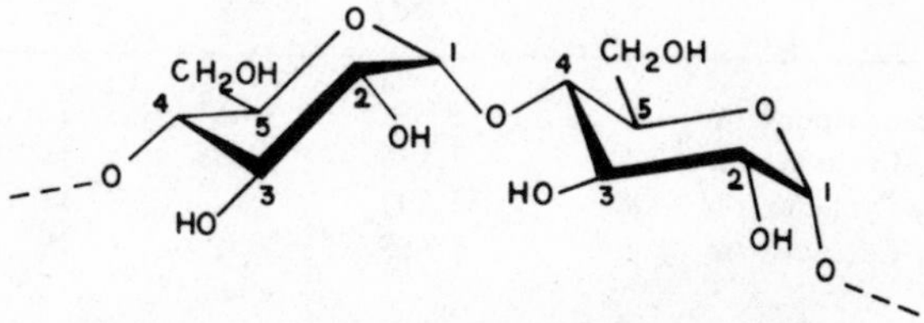

Fig. 1. Schematic diagram of two glucopyranose units of a cyclodextrin molecule illustrating details of the α-(1,4) glycosidic linkage and the numbering system employed to describe the glucopyranose rings

Fig. 2. From left to right, Corey-Pauling-Koltun molecular models of α-, β-, γ-cyclodextrins viewed from the secondary hydroxyl side of the torus

These structures described above are mainly based on X-ray crystallography of the α-cyclodextrin-potassium acetate complex by Hybl et al. [29]. β-cyclodextrin and γ-cyclodextrin are believed to have analogous structures, though X-ray studies on them have not been completed [30—32]. Furthermore, cyclic structures composed of α-(1.4) linked glucopyranose units are consistent with other research:

a) cyclodextrins are non-reducing

b) glucose is the only product of acid hydrolysis of cyclodextrins [33], and 2,3,6-tri-methylglucose is the only product of methylation followed by hydrolysis [34, 35]

c) molecular weights are integral multiples of the value (162.1) for a glucose residue [34]

d) periodate oxidation produces neither formic acid nor formaldehyde [36]

Table 1 lists the dimensional sizes of cyclodextrins as well as some of their important physical properties.

Table 1. Physical properties of the cyclodextrins

Cyclodextrin	Number of glucose residues	Molecular weight (calcu-lated)	Water solubility (g/100 ml)	Specific rotation $[\alpha]_D^{25}$	Cavity dimensions (Å)	
					Internal diameter	Depth
α-Cyclodextrin	6	972	14.5[a]	150.5 ± 0.5[a]	4.5[d]	6.7[d]
β-Cyclodextrin	7	1,135	1.85[a]	162.5 ± 0.5[a]	~7.0[e]	~7.0[e]
γ-Cyclodextrin	8	1,297	23.2[a]	177.4 ± 0.5[a]	~8.5[e]	~7.0[e]
δ-Cyclodextrin	9	1,459	Very soluble[b]	191 ± 3[c]	–	–

[a][19]. [b][3]. [c][21]. [d]From X-ray analysis in [37]. [e]Estimated from Courtald molecular models in [3].

The conformations of cyclodextrins in solution are quite pertinent, since most of the reactions by cyclodextrins are carried out in solution, mostly in water. In general, they are nearly identical with those in the crystalline state. Proton nuclear magnetic resonance (^{1}H–NMR) [38–42], infrared spectroscopy [40, 43], and optical rotatory dispersion spectroscopy [44, 45] demonstrated that the D-glucopyranose units in cyclodextrins are in the C1 chair conformation in dimethyl sulfoxide and D_2O. These data necessarily require that the primary and secondary hydroxyl groups have similar conformation to those in the crystalline state.

A more precise study of the conformation of α-cyclodextrin in solution was made by the 220 MHz ^{1}H-NMR spectroscopy [80]. The chemical shifts and coupling constants were determined by computer simulation of the ^{1}H-NMR spectrum shown in Fig. 3. It was found that on the ^{1}H-NMR time scale all six glucose units have identical conformations and the molecule has hexagonal symmetry. The magnitudes of the vicinal coupling constants ($J_{12} = 3.3$, $J_{23} = 9.8$, $J_{34} = 8.8$, and $J_{45} = 9.6$ Hz) are almost identical with those for its monomer analog, methyl α-D-glucopyranoside ($J_{12} = 3.7$,

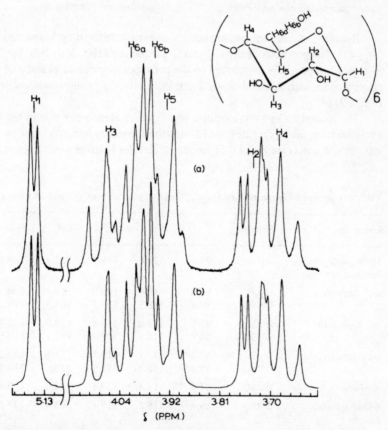

Fig. 3 (a) 220 MHz ^{1}H-NMR spectrum of 0.1 M α-cyclodextrin in D_2O (68 °C, pD = 7.5). (b) computer-simulated spectrum. Reprinted with permission from Wood, D. J., Hruska, F. E., Saenger, W.: J. Amer. Chem. Soc. *99*, 1735 (1977). Copyright by the American Chemical Society

Properties

J_{23} = 9.8, J_{34} = 8.8, and J_{45} = 10.0 Hz), indicating the C1 chair forms of the glucose units in cyclodextrin. However, the J_{56a} (1.8 Hz) and J_{56b} (3.7 Hz) in the cyclodextrin are considerably smaller than those in methyl α-D-glucopyranoside (J_{56a} = 2.3 and J_{56b} = 5.7 Hz), which are between the usual value for the *gauche* hydrogen-hydrogen coupling constant (about 2 Hz) and that for the *trans* hydrogen-hydrogen coupling constant (about 10 Hz). There exist sizable contributions from both the *gauche-gauche* and the *gauche-trans* conformers with respect to the C_5–C_6 bonds in cyclodextrin in aqueous solution. The population of the *gauche-gauche* conformer is slightly larger in the cyclodextrin than in methyl α-D-glucopyranoside.

gauche–gauche conformer   *gauche–trans* conformer

Recently, the full assignments of ^{13}C-chemical shifts of cyclodextrins were made [46–48]. The positions of the C-1 and C-4 carbons shifted downfield by 2–3 and 3–5 ppm, respectively, compared to the corresponding chemical shifts of their linear chain analogs such as α-maltose and amylose, indicating conformational restraints (Table 2) [47].

The secondary hydroxyl groups, which are located in one side of the toruses of cyclodextrins, are hydrogen bonded with the secondary hydroxyl groups of contiguous glucose units (see Fig. 4). These intramolecular hydrogen bondings in cyclodex-

Table 2. Carbon-13 Chemical shifts for α-, β-, and γ-cyclodextrins and related compounds[a, b]

Compound	Solvent	C-1	C-2	C-3	C-4	C-5	C-6
Methyl α-D-gluco-pyranoside	D_2O	92.9	120.6	118.9	122.4	120.9	131.3
α-Cyclodextrin	D_2O	90.5	118.6	120.0	110.7	120.3	131.6
	DMSO	89.7	118.3	119.5	109.5	119.5	131.5
β-Cyclodextrin	D_2O	90.1	118.8	119.8	110.8	120.0	131.6
	DMSO	89.7	118.5	119.2	110.1	119.6	131.6
γ-Cyclodextrin	D_2O	90.3	119.1	119.7	111.5	120.2	131.7
	DMSO	89.9	118.3	118.9	110.6	119.4	131.5
Amylose	DMSO	91.6	118.4	119.8	113.8	120.0	131.2
6-Deoxy-β-cyclo-dextrin	DMSO	89.4	118.5	119.1	103.4	125.1	174.3

[a] From Takeo, K., Hirose, K., Kuge, T.: Chem. Lett. *1973*, 1233.
[b] In ppm upfield from the external CS_2 reference.

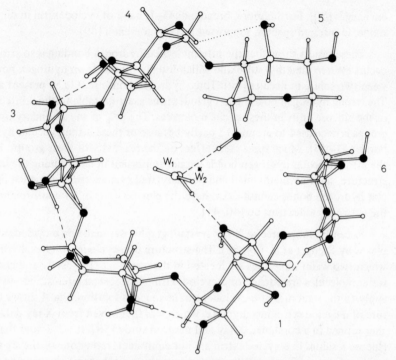

Fig. 4. Structure of α-cyclodextrin hexahydrate; only two water molecules (W_1 and W_2) included in the cyclodextrin cavity are shown. Numbers refer to glucose units. The dotted line between the most rotated glucose 5 and glucose 4 represents a questionable hydrogen bond of 3.36 Å O- - -O distance. Water molecule W_2 is above W_1 and hydrogen bonded to W_1 and to two O(6) hydroxyl groups of α-cyclodextrin. The ∗ mark shows the center of the cyclodextrin molecule. From Saenger, W., Noltemeyer, M., Manor, P. C., Hingerty, B., Klar, B.: Bioorg. Chem. *5*, 187 (1976)

trins were originally proposed by Hybl et al. on the basis of the result of X-ray crystallography [29], and were confirmed by the low ^{1}H-chemical shifts for the secondary hydroxyl groups ($\delta \approx 5.5$ ppm) in dimethyl sulfoxide [49]. Non-anomeric hydroxyls of glucose and most sugars related to glucose, free from the hydrogen bondings, showed their signals in the range 4–5 δ. The study on the hydrogen-deuterium exchanges of cyclodextrins were also favorable for the intramolecular hydrogen bondings. The equilibrium constants for the hydrogen-deuterium exchange reactions

$$2 \text{ CD–OH} + D_2O \rightleftharpoons 2 \text{ CD–OD} + H_2O$$

were determined in dimethyl sulfoxide by ^{1}H-NMR spectroscopy, where CD–OH represents the hydroxyl group of cyclodextrin. The obtained values of the equilibrium constants for the secondary hydroxyl groups of cyclodextrins (0.75 for α-cyclodextrin and 0.65 for β-cyclodextrin) are considerably smaller than that of amylose (0.85). This result indicates that the intramolecular hydrogen bondings of cyclodextrins make the secondary hydroxyl protons more resistant to hydrogen-deuterium exchanges [50]. Furthermore, broadend νO—H band of cyclodextrin in dimethyl sulf-oxide, in infrared spectra, supported this phenomenon [49].

These results show that the intramolecular hydrogen bonding is so strong that it occurs even in dimethyl sulfoxide which usually breaks down hydrogen bonds between two solute molecules [49]. Thus, hydrogen bonds should be present also in H_2O. The strong hydrogen bonding in cyclodextrins is attributable to the limited rotation of the glucose units in these cyclic molecules. The pK_a of the secondary hydroxyl groups is decreased to around 12 partly because of these intramolecular hydrogen bonds [51—53], which has a great effect on the reactivity of these groups. However, the intramolecular hydrogen bonds are not important in determining the macrocyclic structure, since permethylated and peracetylated cyclodextrins, in which intramolecular hydrogen bonds cannot exist, have the glucose units in C1 conformations just as the native cyclodextrins do [40, 41].

Recently, more precise X-ray crystallography was made on α-cyclodextrin complexes by Saenger et al. [54, 55]. The structure of α-cyclodextrin hexahydrate, in which two water molecules are located in the α-cyclodextrin cavity and another four water molecules are outside of α-cyclodextrin, is of special interest, because it may apply to the structure of α-cyclodextrin in aqueous solution. Fig. 4 shows the structure of α-cyclodextrin hexahydrate, which was first solved from X-ray data, and further refined in a combined X-ray and neutron study [56]. It was found that all of six glucose residues in α-cyclodextrin are not equivalent; rather one of the six glucose residues (glucose 5 in Fig. 4) is rotated in such a way that this glucose ring is more nearly normal to the axis of the α-cyclodextrin torus than the other five glucose rings. Thus, the macrocyclic ring in α-cyclodextrin hexahydrate is distorted. The dihedral angles for the five glucoses have values close to the local minimum of the calculated energy, whereas the angles belonging to the rotated glucose (number 5) indicate considerable steric strain. The ring of interglucosidic, intramolecular hydrogen bonds is broken at this glucose with $O(2)...O(3)$ distances of 4.66 and 3.35 Å. The $C(6)$—$O(6)$ bonds for this (number 5) and for another glucose (number 1) are in the (+)*gauche* orientation to allow hydrogen bonding to one of the included water molecules. The second water molecule is hydrogen bonded to this first one (and then to an adjacent cyclodextrin molecule). The two water molecules are not located on the toroidal axis of the cyclodextrin but displaced by 0.6 Å from this axis. The distortion of the macrocyclic ring of α-cyclodextrin in water was also proposed on the basis of optical rotation measurements [57]. Interestingly, the distortion of the macrocyclic ring of α-cyclodextrin vanishes when a-cyclodextrin forms inclusion complexes with guest compounds [54, 55]. The relation between the conformational change in cyclodextrins and their inclusion complexes will be described in the next Chapter.

α-,β-, and γ-Cyclodextrins have higher free energy than the corresponding linear glucose chains by +2.29, +1.71, and +1.99 kcal/mole at 25 °C [58, 59]. Thus, cyclizations of linear glucose chains to cyclodextrins are energetically unfavorable. In this process, the enthalpy increases +6.60, +4.41, and +4.40 kcal/mole, respectively, for α-, β-, and γ-cyclodextrins which show the extent of energetic instability introduced by the cyclization. On the other hand, the entropy terms (+14.4, +9.1, and +8.1 e. u., respectively, for α-, β-, and γ-cyclodextrins) are favorable for the cyclization process, which probably reflects the reorganization of water structure around the glucose chain on the cyclization [59].

Cyclodextrin molecules are fairly stable in alkaline solution. However, they are quite susceptible to acid catalysis. For example, the rate constants of hydrolysis of β-cyclodextrin at pH − 0.133, 40 °C and 100 °C, respectively, are 1.0×10^{-5} and 4.8×10^{-2} min^{-1}, which correspond to the half-lives of 48 days and 14 min. The activation energy for the scission of glycosidic bonds in β-cyclodextrin is 34.2 kcal/mole, whereas that in the case of maltose, a linear sugar, is only 30.5 kcal/mole. On the other hand, the activation entropy of the hydrolysis of β-cyclodextrin (16.6 e. u.) is larger than that of maltose (7.3 e. u.). Thus, in the temperature range of 20−100° β-cyclodextrin is hydrolyzed by acid 2−3 fold more slowly than maltose is [60]. Because of the larger rate constant of hydrolysis of linear sugar than cyclodextrin, the first-order rate constant for the acid hydrolysis of cyclodextrin, determined from reducing power data, increases during the reaction [60−62]. Taka-amylase A markedly catalyzes the hydrolyses of cyclodextrins [63, 64].

III. Inclusion Complex Formation

1. Detection of Complex Formation and Structure of the Complexes

One of the most important characteristics of cyclodextrins is the formation of inclusion complexes with various compounds (guests), in which guest compounds are included in the cavity of cyclodextrins (host) [20, 65–67]. Guest compounds range from polar reagents such as acids, amines, small ions such as ClO_4^-, SCN^-, and halogen anions [68] to highly apolar aliphatic and aromatic hydrocarbons and even rare gases [69]. Inclusion complexes can be formed either in solution or in the crystalline state. Water is usually used as solvent, although inclusion complex formation also takes place in dimethyl sulfoxide and in dimethyl formamide [70].

The molar ratio of guest to host (cyclodextrins and their derivatives [71, 72]) is usually 1 : 1 in inclusion complexes formed in solution, with the exceptions of the inclusion complexes of cyclodextrins with long chain aliphatic carboxylic acids [73], methyl orange [74], and certain barbituric acid derivatives [75].

Reaction of diamines such as hexamethylenediamine included in β-cyclodextrin with terephthaloyl chloride gave inclusion polyamides, in which polyamide chains pass through a tunnel formed by the cavities of the cyclodextrins [76].

Within a similar series of substrates, the tendency to complex with cyclodextrins can be qualitatively correlated with the fit of the substrate to the cavity of the cyclodextrin [77]. However, when comparing the tendencies of complexing of widely different substrates, the fit can not be satisfactorily predictive.

Inclusion complex formation can be detected by variety of spectroscopic methods; e.g. nuclear magnetic resonance, absorption, fluorescence, optical rotation etc.

The most direct evidence for the inclusion of a guest into the cyclodextrin cavity in solution was obtained by proton nuclear magnetic resonance (^{1}H-NMR) spectroscopy [78]. The H-3 and H-5 atoms of α-cyclodextrin, which are directed toward the interior of the cyclodextrin cavity, showed a significant upfield shift upon addition of substituted benzoic acids to cyclodextrin solutions in D_2O. On the other hand, the H-1, H-2, and H-4 atoms, located on the exterior of the cavity, showed only a marginal upfield shift. Obviously, the large upfield shift for the H-3 and H-5 atoms can be ascribed to an anisotropic shielding effect of the benzene rings of the benzoic acids included in the cyclodextrin cavity. The penetration of the benzene ring of phenobarbital (1) into the cyclodextrin cavity seems to be shallower than that of the ben-

10

zene rings of the benzoic acids: since the upfield shift of the H-3 atom of cyclodextrin is small irrespective of the large upfield shift of the H-5 atom (Table 3) [79].

(1)

In inclusion complex formation of p-iodoaniline with α-cyclodextrin, the H-3 protons showed an upfield shift, while the H-5 protons showed a downfield shift. The chemical shifts of the H-1, H-2, and H-4 protons were unaffected. These results also indicate the inclusion of p-iodoaniline in the cavity of α-cyclodextrin. The upfield shift for the H-3 atom was attributed to the anisotropic shielding effect of the benzene ring of p-iodoaniline, which is included in the cyclodextrin cavity. The downfield shift for the H-5 atom was interpreted in terms of the van der Waals deshielding effect of the iodine atom of p-iodoaniline as well as the ring-current effect of the benzene ring [80].

Qualitatively similar results were obtained for the inclusion complex of β-cyclodextrin with tranquilizing drugs such as the phenothiazines (2). On complex formation, the protons located in the cavity of β-cyclodextrin are subject to anisotropic shielding, and thus appear at higher magnetic field, whereas protons of the phenyl and N-substituted groups in the drugs are shifted toward a lower magnetic field. The aromatic portion of the drug is included in the cyclodextrin cavity [81].

(2)

Table 3. Changes in proton chemical shifts of β-cyclodextrin on complex formation with substrate[a]

Substrate	Chemical shift (ppm)[b]					
	H-1	H-2	H-3	H-4	H-5	H-6
Benzoic acid[c]	+ 0.04	+ 0.04	+ 0.16	+ 0.03	+ 0.19	+ 0.05
p-Hydroxybenzoic acid[c]	+ 0.04	+ 0.04	+ 0.14	+ 0.04	+ 0.21	+ 0.06
Phenobarbital (1)[d]	+ 0.04	+ 0.03	0.00	+ 0.06	+ 0.31	+ 0.11

[a] H-3 and H-5 are directed toward the interior of the cavity, whereas H-1, H-2 and H-4 are located on the exterior (see Fig. 1).
[b] Chemical shifts upon saturation of a 2% aqueous solution of β-cyclodextrin with substrate; the + sign refers to a shift to higher magnetic field.
[c] From Demarco, P. V., Thakkar, A. L.: Chem. Commun. 1970, 2.
[d] From Thakkar, A. L., Demarco, P. V.: J. Pharm. Sci. 60, 652 (1971).

The molecular dispositions of guest compounds in the cavities of cyclodextrins were examined by use of ^{1}H-NMR and ^{13}C-NMR spectroscopy [82–84]. On complexation of sodium p-nitrophenolate with α-cyclodextrin in aqueous solution, only the H-3 proton showed a shift toward higher magnetic field, the H-5 proton showing little shift. On the other hand, the *meta*-proton of the guest compound was deshielded by 0.30 ppm whereas the *ortho*-proton only by 0.16 ppm [82]. In the ^{13}C-NMR spectra, the *meta*-carbons of the guest compound were deshielded by 1.43 ppm whereas the *ortho*-carbons were shielded by 0.28 ppm [83]. Furthermore, in a proton homonuclear Overhauser experiment, irradiation of the H-3 resonance of α-cyclodextrin produced a larger enhancement of the p-nitrophenolate resonances than irradiation of any other protons. Besides, a large enhancement was observed for the *meta*-proton of the guest compound whereas there was no significant effect on the intensity of the *ortho*-proton resonance [82]. These facts indicate that this guest compound penetrates the α-cyclodextrin cavity from the secondary hydroxyl side with the nitro group first, and only to the extent that the *meta*-protons are in close proximity to the cyclodextrin H-3 protons.

The conclusion that both p-nitrophenol and p-nitrophenolate penetrate the cavity with the nitro group first was confirmed by comparing the dissociation constant (K_d) of the complexes of these compounds with those of methyl substituted derivatives [Table 4]. Introduction of methyl groups at the 2 and 6 positions of p-nitrophenol weakens binding only slightly by a factor of about 2.4. However, introduction of a single methyl group at the 3 position of p-nitrophenol weakens the binding by a factor of about 100, whereas introduction of methyl groups at both the 3 and 5 positions completely inhibits binding. Obviously, the steric hindrance around the nitro group has a marked effect on the binding [84a].

Furthermore, the sign of the induced Cotton effect supported inclusion of sodium p-nitrophenolate in the cavity of cyclodextrin with its axis parallel to the axis of the cyclodextrin cavity [84 c]. The induced Cotton effect due to the [1]L$_a$ transition (moment), polarized along the sodium p-nitrophenolate, showed a positive Cotton effect, which is consistent with theory predicting that an electric dipole moment on the axis of the cyclodextrin ring gives a positive Cotton effect, whereas one perpendicular to the axis gives a negative one [101].

^{1}H-NMR spectroscopy and [1]H-homonuclear Overhauser experiments were also carried out to examine the molecular dispositions of benzoic acid and sodium benzoate in the cavity of α-cyclodextrin [84b]. For both benzoic acid and sodium benzoate, complexation with α-cyclodextrin showed shifts of the *ortho*- and *meta*-protons toward lower magnetic field. The *ortho*-protons exhibit about twice the change in chemical shift as the *meta*-protons. Furthermore, in [1]H-homonuclear Overhauser experiments, the *ortho*-protons showed a large enhancement, whereas the *meta*-protons remained virtually unchanged. From these results, it was proposed that both benzoic acid and sodium benzoate are included in the cyclodextrin cavity in such a way that the carboxyl group enters first and the benzene ring last. However, the structure of the complex with the carboxylate ion inside the cavity is surprising, since the hydrophilic carboxylate ion usually prefers to be in aqueous media instead of in the apolar cyclodextrin cavity. If the proposed structures of the complexes are correct, it will require a complete reconsideration of the binding forces of cyclodextrins.

12

Table 4. The dissociation constants (K_d) of α-Cyclodextrin inclusion complexes at 25° [a]

Guest compound	K_d (10^{-2} M) [b]
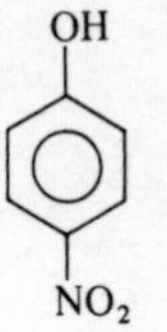 OH / benzene / NO$_2$ (p-nitrophenol)	5.3
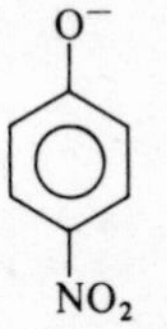 O$^-$ / benzene / NO$_2$ (p-nitrophenolate)	0.040
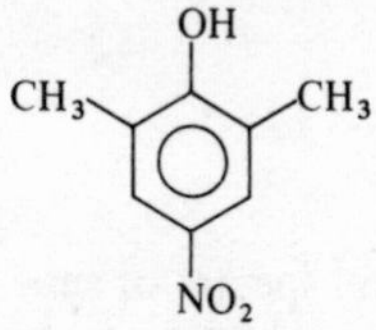 CH$_3$ / OH / CH$_3$ / benzene / NO$_2$	~~0.18~~ *18.0*
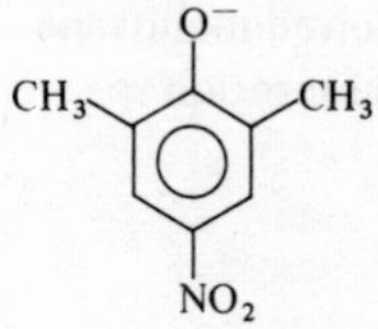 CH$_3$ / O$^-$ / CH$_3$ / benzene / NO$_2$	0.0094
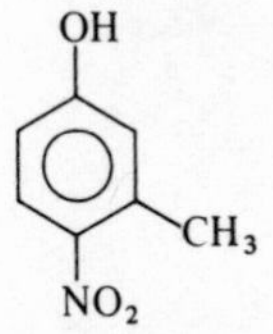 OH / benzene / CH$_3$ / NO$_2$	N. B. [c]
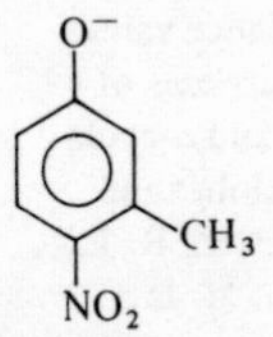 O$^-$ / benzene / CH$_3$ / NO$_2$	4.2 *weaker binding here*

Table 4 (continued)

Guest compound	K_d $(10^{-2}$ M$)^b$
OH / CH$_3$ · · · CH$_3$ / NO$_2$	N. B.[c]
O$^-$ / CH$_3$ · · · CH$_3$ / NO$_2$	N. B.[c]

[a] From Bergeron, R. J., Channing, M. A., Gibeily, G. J., Pillor, D. M.: J. Amer. Chem. Soc. *99*, 5146 (1977).
[b] Determined by spectrophotometric methods.
[c] N. B. = no binding.

Inclusion complex formation often causes changes in absorption spectra of sub-strates [74, 85–90]. For example, Fig. 5 shows a typical spectral change [74]. Addition of α-cyclodextrin to an aqueous solution of *p-t*-butylphenol causes a spectral change almost identical with that observed when the phenol is dissolved in dioxane. Glucose, which has no cavity, showed no significant effect on the absorption spec-

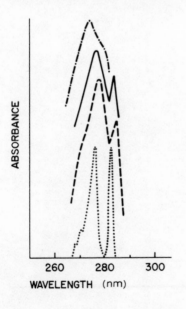

Fig. 5. Ultraviolet absorption spectrum of *p-t*-butyl-phenol in various solvents. The absorbance values are aribitrarily shifted vertically for purposes of clarity. — · — · — water; ——— water and α-cyclo-dextrin; — — — dioxane; · · · · · · cyclohexane. Reprinted with permission from VanEtten, R. L., Sebastian, J. F., Clowes, G. A., Bender, M. L.: J. Amer. Chem. Soc. *89*, 3242 (1967). Copyright by the American Chemical Society

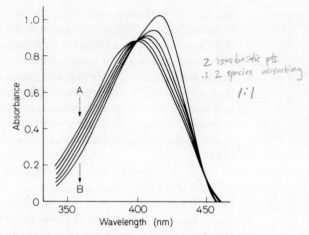

Fig. 6. Spectrum of *p*-nitrophenol at varying α-cyclodextrin concentrations; pH 11.0, 20 °C. The cyclodextrin concentrations are changed from 0 to 10^{-2} M from A to B. The concentration of *p*-nitrophenol is 5×10^{-5} M. Isosbestic points are observed at 398 and 446 nm. Reprinted with permission from Cramer, F., Saenger, W., Spatz, H.-Ch.: J. Amer. Chem. Soc. *89*, 14 (1967). Copyright by the American Chemical Society

trum. Analogous results were obtained for cyclodextrin complexes of iodine and N-acetyltyrosine ethyl ester [3]. The similarity of the aqueous cyclodextrin solution spectrum to the dioxane solution spectrum strongly shows that the chromophore is included in the cyclodextrin cavity, which has an apolar atmosphere like ether as mentioned in Chap. II. In most cases, the stoichiometries of inclusion complexes are shown to be 1:1 by isosbestic points in spectrophotometric titrations (Fig. 6) [87].

Inclusion of substrates in the cyclodextrin cavity is also supported by fluorescence measurements [87, 91]. Addition of 0.01 M β- and γ-cyclodextrins to an aqueous solution of 1-anilino-8-naphthalenesulfonate (ANS) enhanced the fluorescence of ANS by 10 fold (Fig. 7). This result is consistent with the transfer of ANS from the aqueous medium to the apolar cyclodextrin cavity, since ANS exhibits only a very weak fluorescence in water, but shows strong fluorescence in organic solvents such as ethyl alco-

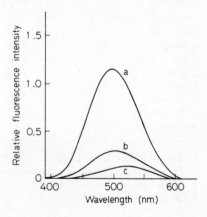

Fig. 7. Fluorescence spectrum of 1-anilino-8-naphthalenesulfonate (ANS); (a) 10^{-4} M ANS + 10^{-2} M β-cyclodextrin, (b) 10^{-4} M ANS + 10^{-2} M α-cyclodextrin, and (c) 10^{-4} M ANS. Reprinted with permission from Cramer, F., Saenger, W., Spatz, H.-Ch.: J. Amer. Chem. Soc. *89*, 14 (1967). Copyright by the American Chemical Society

hol. On the other hand, α-cyclodextrin exhibited a much smaller effect on the fluorescence of ANS, which is attributable to the inability of the small cavity of α-cyclodextrin to completely enclose the naphthalenesulfonate residue [87]. A similar fluorescence enhancement was observed on complexation of cyclodextrin with 6-p-toluidinylnaphthalene-2-sulfonate [92] and dansyl (1-dimethylaminonaphthalene-5-sulfonyl) derivatives [93–96]. The pronounced fluorescence enhancement accompanying inclusion complex formation with cyclodextrins was used to monitor the rate of hydrolysis of cyclodextrins [92, 96]. Furthermore, spraying of cyclodextrins on thin-layer chromatograms enhanced the fluorescence of compounds, facilitating their detection [95].

Since cyclodextrins have an assymmetric cavity, inclusion complex formation of optically active substrates such as nucleic acids change their circular dichroism spectra [97, 98]. Furthermore, circular dichroism is induced on binding of optically inactive compounds to cyclodextrins [86, 99–105]. The substrate may be fixed in one preferable dissymmetric configuration in the cavity of the cyclodextrin.

Inclusion complex formation can be also detected by titration. The pK_a of acids in the presence of cyclodextrins are larger than (or equal to) those in its absence, whereas the pK_a of phenols in the presence of cyclodextrin are smaller than (or equal to) those in its absence (Table 5) [106]. For example, p-nitrophenol ($\Delta pK_a = -0.94$). can be compared with p-nitrobenzoic acid ($\Delta pK_a = +0.79$). This result shows that p-nitrophenolate ion penetrates into the cyclodextrin cavity more easily than unionized p-nitrophenol, in contrast to the reverse situation for the benzoic acid. The unusual result with p-nitrophenol (the anionic species binds with cyclodextrin more strongly than the neutral species) was interpreted in terms of an extensive charge delocalization in the p-nitrophenolate ion. Furthermore, p-nitrophenol itself is a good hydrogen-bonding donor, and its affinity with the solvent (water) molecules may destabilize its complex with the cyclodextrin.

Inclusion complex formation can also be followed by electron spin resonance spectroscopy, using spin-labeled substrates such as 2,2,6,6-tetramethyl-4-oxopiperi-

Table 5. Effects of cyclodextrins on the dissociation constants of acids and phenols at 25° [a]

Compound	pK_a' [b]	pK_a [c]	ΔpK_a [d]
Acetic acid	4.83	4.76	+ 0.07
Benzoic acid	5.20	4.11	+ 1.09
Nicotinic acid	4.85	4.85	0.00
Cinnamic acid	5.80	4.43	+ 1.37
Phenol	9.81	9.81	0.00
o-Nitrophenol	7.21	7.21	0.00
m-Nitrophenol	8.00	8.29	− 0.29
p-Nitrophenol	6.15	7.09	− 0.94

[a] From Connors, K. A., Lipari, J. M.: J. Pharm. Sci. 65, 379 (1976).
[b] In the presence of 0.02 M α-cyclodextrin.
[c] In the absence of α-cyclodextrin.
[d] $\Delta pK_a = pK_a' - pK_a$.

dinyl-1-oxy (3) [107–111]. On binding of (3) to β-cyclodextrin, the isotropic nitrogen hyperfine coupling constant decreased, indicating the movement of (3) to an environment less polar than water. The rotational correlation time of (3) increased from 1.7×10^{-11} sec to 2.8×10^{-10} sec in the presence of 15.82×10^{-3} M β-cyclodextrin, which is consistent with partial restriction of molecular rotation of (3) on inclusion complex formation [110].

(3)

Polarography can be also a probe to the complex formation of cyclodextrin [112, 113]. For example, formation of clathrate compound of hydroperoxide with β-cyclodextrin can be followed by the decrease of its half wave potential with an increasing concentration of the cyclodextrin [112].

Inclusion complexes in solution are not static species. Rather, substrates included in the cavity rapidly exchange with free substrate molecules. They can also have different orientations about the axis parallel to the cavity and/or are rapidly spinning about this axis within the cavity. This was shown by only one time-averaged signal for each of the protons of cyclodextrin in the presence of benzoic acids in D_2O solution [78].

The rate constant of recombination of the substrate with cyclodextrin (k_R) and that of dissociation (k_D) were measured by a temperature-jump technique [87] or an ultrasonic relaxation technique [114] (Table 6). The k_R of all the guest compounds

Table 6. Recombination rate constants (k_R) and dissociation rate constants (k_D) of the complex between cyclodextrin and the guest compounds

Guest compound	k_R $(10^9 \text{ M}^{-1} \text{ sec}^{-1})$	k_D (10^6 sec^{-1})	K_d [a] (10^{-2} M)
p-Nitrophenol[b]	≥ 0.04	≥ 0.1	0.26
p-Nitrophenolate[b]	0.14	0.031	0.027
I^- [c]	0.065	3.6	5.5
SCN^- [c]	0.044	4.4	10
Br^- [c]	0.045	6.9	15
NO_3^- [c]	0.045	8.2	18
Cl^- [c]	0.054	21	39
ClO_4^- [c]	2.0	74	3.7

[a] Dissociation constant of the complex determined by spectrophotometric method.
[b] Determined at 14 °C by temperature-jump technique in Cramer, F., Saenger, W., Spatz, H.-Ch.: J. Amer. Chem. Soc. 89, 14 (1967); α-cyclodextrin was used as host.
[c] Determined at 25 °C by ultrasonic relaxation method in Rohrbach, R. P., Rodriguez, L. J., Eyring, E. M., Wojcik, J. F.: J. Phys. Chem. 81, 944 (1977); β-cyclodextrin was used as host.

17

except that for ClO_4^- are large but considerably smaller than diffusion-controlled. This result gives information about the rate-determing step of recombination, about which several proposals were made. One proposal is that a special steric requirement in the recombination step in the inclusion complex formation suppresses the k_R values considerably below the diffusion-controlled limit, though the recombination is really diffusion-controlled [87]. The second proposal is that the breakdown of the water structure inside the cyclodextrin cavity and removal of some water molecules out of the cavity are rate-determining [87]. The third proposal ascribes the rate-determining step as a conformational change of cyclodextrin [114]. As mentioned later in this Chapter, a conformational change of cyclodextrin in its inclusion complex formation with guests has been proposed on the basis of the results of X-ray crystallography [54—56, 120, 125]. At the present time, it is hard to determine which proposal is right.

ClO_4^- ion is too large to be accomodated in the cyclodextrin cavity, and probably forms a straddle type complex. Thus, it is quite reasonable that its k_R is diffusion-controlled.

k_D increases with decreasing stability of the complex (larger K_d) in a series of anions (I^-, SCN^-, Br^-, NO_3^-, and Cl^-), whereas the k_R of these ions are almost equal to one another.

Recently, molecular motions in inclusion complexes of α-cyclodextrin with p-methylcinnamate, m-methylcinnamate, and p-t-butylphenolate ions were investigated by ^2H and ^{13}C nuclear relaxation [115, 116]. The overall tumbling motion of the three substrates were decelerated by inclusion complex formation by about 4 fold. The overall reorientation time for the cyclodextrin in the complex depended on the substrate. The internal rotations of methyl groups of p-methylcinnamate and m-methylcinnamate anions were hindered by inclusion complex formation, showing that these methyl groups are located inside or at least in contact with the macrocycle. In contrast, the bulky t-butyl group in the cyclodextrin-p-t-butylphenolate ion complex is probably located outside the cavity, since its internal motion is affected only a little on complexation. It was proposed that dynamic rigidity, defined by the coupling between the molecular motion of the cyclodextrin and the guest compound, is important.

In contrast to 1 : 1 inclusion complexes in solution, the ratios of guest to host are usually nonstoichiometric in the crystalline state. This is associated with the three dimensional structure of crystalline inclusion complexes [29, 31]. X-ray crystallography on many inclusion complexes has been carried out [29—31, 37, 54—56, 117—131]. There are two different forms of inclusion complexes in the crystalline state, namely, channel and cage-type structures [118]. Channel structures develop when cyclodextrins stack on top of one another to yield endless channels, in which the guest molecules are included. Cage structures are a result of a displaced arrangement of cyclodextrins, in which guest molecules are located in discrete, small cavities represented by annular apertures.

In the α-cyclodextrin-iodine tetrahydrate, the iodine molecule is included in the cavity and is co-axial with the α-cyclodextrin. The I–I distance is a conventional 2.677 Å. The distances between the one I atom of the iodine molecule and C(5) and C(6) atoms are 3.75—4.17 Å, which are close to van der Waals contact (the sum of

18

the van der Waals radius of iodine (2.15 Å) and that of methylene carbon (2.00 Å) is 4.15 Å). The separations between the other I atom of the same iodine molecule and the O(4) atom exceed contact separations. The molecules are arranged in herring-bone "cage-type" fashion, with the four water molecules as space-filling mediators; the structure is held together by an intricate network of hydrogen bonds [119].

The 2:1 complex of methyl orange sodium salt or potassium salt with α-cyclo-dextrin in the crystalline state has a typical channel structure built up by the stacking of α-cyclodextrins. The methyl orange anions are located in the channel formed by the cyclodextrin molecules. The azo group and the benzene ring are accommodated in the cyclodextrin cavity, whereas the dimethylamino and sulfonate groups protrude from the cavity and are in contact with the adjacent α-cyclodextrin rings. The azo group is located at the "neck" of the cavity. The sulfonate group is hydrogen-bonded to the primary hydroxyl groups of α-cyclodextrin. No significant difference of the structure was observed between in the α-cyclodextrin complex with methyl orange sodium salt and in the α-cyclodextrin complex with its potassium salt [126].

The X-ray crystallography on the cyclodextrin complex with mono- or di-substi-tuted benzene is very significant, since the reactions catalyzed by cyclodextrins (des-cribed in detail in the following part of this book) mostly use these kinds of com-pounds as substrates. Thus, the information about the structures of the complexes is required in connection with their reactivity. In the α-cyclodextrin-p-iodoaniline trihydrate, the p-iodoaniline molecule is located within the α-cyclodextrin cavity so that the hydrophobic I atom is in the hydrophobic pocket and the amino group pro-trudes from the O(2), O(3) side of the ring by about 1.6 Å [121, 130]. No consider-able distortion of the structure of p-iodoaniline on the complex formation was ob-served. However, the benzene ring of the p-iodoaniline molecule quite tightly fits the cavity. Thus, the distances between the two H atoms of the guest *ortho* to the I atom and the O(4) atoms of the adjacent glucose units are 2.286 and 2.333 Å, both of which are shorter than the ideal van der Waals distance (2.6 Å). The distance between the amino group and the secondary hydroxyl groups are greater than 4.0 Å. The sub-strate molecule is not hydrogen bonded to the O(6) hydroxyl groups of the host α-cyclodextrin molecule.

In the α-cyclodextrin-sodium benzenesulfonate complex, the benzene ring is located in the cavity, whereas the sulfonate group protrudes from the O(2), O(3) side (the secondary hydroxyl group side) of the cavity (and is hydrogen-bonded to the primary hydroxyl groups of the adjacent α-cyclodextrin molecule) [127].

Mostly the C(6)−O(6) bonds are preferentially directed away from the center of the α-cyclodextrin molecule (the O(5)−C(5)−C(6)−O(6) angles are in the (−)-*gauche* range). However, in some complexes where hydrogen bonding to the enclosed substrate is possible, e.g. with alcohols, one or two of the C(6)−O(6) bonds point toward the center of the cavity to allow the formation of the hydrogen bonds [122, 125]. In the latter cases, the angle about the C(5)−C(6) bond is in the (+)*gauche* range.

An important finding of X-ray crystallography is the conformational change of a cyclodextrin accompanying inclusion complex formation [54−56, 120, 125]. In a water inclusion complex of α-cyclodextrin, which can represent the empty cyclo-dextrin molecule before complex formation, the macrocyclic conformation of cyclodextrin is unsymmetrical. One of the six glucose units is rotated and the macrocyclic ring allegedly collapsed to fit the 5 Å wide cavity to the 3.8 Å thick water molecules [54–56] (see page 8). However, cyclodextrins included with guest compounds have an almost purely cyclic form [54–56, 119–121, 127, 130]. The dihedral angles about $C(4)–O(4)–C(1')$ glucosidic bond, Φ and Ψ, are about $+165$ and $-168°$ [132] which corresponds closely to the global minimum derived from potential energy calculations for the rotation of vicinal glucoses about the bonds [133, 134].

As mentioned above, the substrates which can sufficiently fill the cyclodextrin cavity occupy the center of the cyclodextrin cavity. However, the substrates which are too small to fill the cyclodextrin cavity are statistically disordered. For example, only one methanol molecule is enclosed within the α-cyclodextrin cavity but it is distributed over two sites with equal (50%) population. One site (50% population) is near the $O(6)$ side of the cavity and here methanol is hydrogen-bonded with $O(6)$ hydroxyl group; the other 50% within the cavity and not involved in hydrogen bonding [125]. This is also the case for krypton [123] and n-propanol [122]. On the basis of this finding, Saenger et al. proposed that the release of strain energy of cyclodextrins (change from the high energy conformation of the cyclodextrin-water complex to the lower energy conformation of the cyclodextrin-guest complex) is the driving force of inclusion complex formation [54–56, 120, 125]. This will be discussed in detail later in this Chapter.

In addition to inclusion complex formation where the substrate is included in the cavity, cyclodextrins can form coordination complexes with Cu^{2+} using the many hydroxyl groups of the cyclodextrins [135, 136]. Two pairs of C_2 and C_3 secondary hydroxyl groups of contiguous glucose units are cross-linked by the $Cu(OH^-)_2 Cu$ ion bridge in the α-cyclodextrin coordination complex (4) and by $Cu(OH^-)(O^{2-})Cu$ ion bridge in the β-cyclodextrin coordination complex (5).

4

5

β-Cyclodextrin binds to the active center of pancreatic amylase in a molar ratio of $3:1$. Here, cyclodextrin functions as a competitive inhibitor, not as a host. Cyclodextrins are accomodated in a trough of the amylase molecule, where the substrate, amylose, binds [137–139]. Cyclodextrins are also extremely potent inhibitors of *Aerobacter aerogenes* pullulanase [140].

Selective binding of cyclodextrins with substrates can be used for chlomatography [141–144]. Resins containing cyclodextrins can effectively separate some nucleic acids. Cyclodextrin stationary phases are useful for fractionation of substrates.

which can bind in the cyclodextrin cavity [141 a]. A Sepharose column, incubated with α-cyclodextrin, can separate β-amylase from albumin, though the same column, incubated in the absence of α-cyclodextrin, can not [144]. Epichlorohydrin cross-linked cyclodextrin gel can be used for partial resolution of diastereomers of CrATP [141 b].

2. The Determination of Dissociation Constants of Inclusion Complexes

The most widely used method to determine the dissociation constant (K_d) of the inclusion complex (S · C) between cyclodextrin (C) and substrate (S) shown in equation 1 is absorption spectroscopy [74, 87–90, 145]. The values of K_d can be obtained from the observed change in absorbance and the added concentration of cyclodextrin according to the Benesi-Hildebrand method [146].

$$C + S \underset{K_d}{\rightleftharpoons} S \cdot C \tag{1}$$

Instead of absorbance change, change in ^{1}H-chemical shift, ^{13}C-chemical shift [116], fluorescence intensity [87, 147, 148], optical rotation [70, 98], conductance [70], the wave height in polarography [113], etc. can be used.

Another important source of K_d values is the kinetic method. As mentioned in the following Chapters of this book cyclodextrins accelerate or decelerate various kinds of reactions. The observed reaction rate in the presence of cyclodextrins is a weighted average of the rate of reaction of the free substrate and the rate of reaction of the substrate included in the cyclodextrin. Thus, the K_d value can be obtained from the dependence of the observed rate on the added concentration of cyclodextrin. Furthermore, the K_d values for unreactive guest compounds can be determined from the dependence of the reaction rate of a reactive substrate in the presence of a fixed amount of cyclodextrin with varying concentrations of the added unreactive guest. The unreactive guest inhibits the reaction of reactive substrate by competitive binding in the cyclodextrin cavity [74].

The solubility method was often used to get K_d values before spectroscopy could be easily utilized. This method takes advantage of the change of solubility of substrates in aqueous solution induced by inclusion complex formation. In general, the solubility of cyclodextrins is considerably reduced by inclusion complex formation [19]. On the other hand, the solubility of a substrate on complex formation is generally increased but depends on the substrate and cyclodextrin. For example, the solubilities of n-aliphatic, benzoic, and iodobenzoic acids were increased in the presence of α- and β-cyclodextrins, whereas the solubilities of sterically bulky acids such as 2,3,5,6-tetramethybenzoic acid were hardly affected by cyclodextrins [73]. In some cases, cyclodextrins decrease the solubility of organic molecules. Plots of solubility of substrate versus the concentration of cyclodextrins can be related to the dissociation constant for the cyclodextrin-substrate complexes if the stiochiometries of the complexes are known [149–154]. However, this method makes it so difficult to obtain accurate dissociation constants for inclusion complexes; therefore, spectrophotometric methods are usually employed now.

Table 7 shows some of the values of the dissociation constants determined by spectrophotometric methods as well as those obtained by kinetic methods. In Table 7, the enthalpy change (ΔH) and entropy change (ΔS) for the formation of inclusion complexes, determined from the temperature dependence of K_d, are also shown. The thermodynamic parameters for inclusion complex formation can be also determined directly by calorimetry [155, 156].

Lewis and Hansen determined precisely the thermodynamic parameters for complex formation of α- and β-cyclodextrins with variety of guests by calorimetry [156]. The guest compounds examined are phenols, benzoic acids, indole, hydrocinnamic

Table 7. Thermodynamic parameters in the formation of inclusion complexes

Substrate	Cyclo-dextrin[a]	K_d [b] $(10^{-3}$ M)	ΔH (kcal/mol)	ΔS (e. u.)	Method of determination[c]	References
Phenol	α-CD	53	–	–	S	[74]
3,5-Dimethylphenol	α-CD	16	–	–	S	[74]
p-Nitrophenol	α-CD	2.6[d]	–4.2	–2.8	S	[87]
p-Nitrophenolate ion	α-CD	0.27[d]	–7.2	–8.7	S	[87]
	α-CD	–	–9.04	–14.97	C	[155]
	β-CD	–	–3.40	+1.21	C	[155]
Phenyl acetate	α-CD	22	–	–	K	[74]
m-Chlorophenyl acetate	α-CD	5.6	–	–	K	[74]
	β-CD	3.5	–1	+8	K	[74]
	β-CD	4.7	–	–	S	[74]
m-Ethylphenyl acetate	α-CD	11	–	–	K	[74]
	β-CD	2.2	–4.6	–3	K	[74]
3,4,5,-Trimethylphenyl acetate	β-CD	5.0	–2.5	+2	K	[74]
p-Chlorocinnamate	α-CD	5.1	–	–	S	[74]
	α-CD	5.1	–	–	K	[74]
Benzoylacetic acid	β-CD	9.8[e]	–5.7	–8.6	K	[158]
p-Methylbenzoylacetic acid	β-CD	4.7[e]	–6.6	–9.8	K	[158]
m-Chlorobenzoylacetic acid	β-CD	6.0[e]	–5.2	–6.0	K	[158]
Diisopropyl phosphorofluoridate	α-CD	210	–7.3	–21	K	[159]

[a] α-CD and β-CD represent α-cyclodextrin and β-cyclodextrin.
[b]   Values at 25° unless otherwise noted.
[c]   S, K, and C, respectively, refer to spectrophotometric methods, kinetic methods, and calorimetry.
[d]   14°.
[e]   50.3°.

acid, L-phenylalanine, L-tyrosine, L-tryptophan, anilinium perchlorate, perchloric acid, sodium perchlorate, L-mandelic acid, acetic acid, and pyridine. They found that the compensation rule between ΔH and ΔS holds with an isoequilibrium temperature of 265 °K for these substrates. This result indicates an important role of water in inclusion complex formation, since the compensation effect has been observed often for reactions in water.

The compensation between ΔH and ΔS was also observed for inclusion complex formation of β-cyclodextrin with antiinflammatory drugs such as fenamates [86] and with barbituric acid derivatives [157], where the isoequilibrium temperatures are 284 and 372 °K, respectively.

Calorimetry indicated that the α-cyclodextrin-perchlorate complexes also include the cation [156]. Anilinium perchlorate, perchloric acid, and sodium perchlorate have significantly different enthalpy and entropy changes for the complex formation with α-cyclodextrin ($\Delta H = -12.3, -7.5$, and -9.7 kcal/mole, respectively, while $\Delta S = -35, -17$, and -23 e.u., respectively). Na^+ apparently does not bind to α-cyclodextrin, since sodium chloride did not show any interactions with α-cyclodextrin. Thus, the only reasonable explanation is that α-cyclodextrin is binding the ion pair or that α-cyclodextrin binds ClO_4^- and this complex in turn binds the cation.

3. Binding Force of the Complexes

The nature of the binding force still remains controversial. The interaction force for inclusion complex formation can not be classical apolar binding such as that for enzyme-substrate complex formation, since inclusion complex formation is associated with a favorable enthalpy change and an unfavorable (or slightly favorable) entropy change (Table 7). Usually apolar binding is characterized by a very favorable entropy change [160–162].

Several proposals were made to interpret this sizably favorable enthalpy change [2, 7]:

a) Van der Waals interactions between guest and host [66, 74, 84, 89]

b) hydrogen bonding between the guest and the hydroxyl groups of cyclodextrin [65, 159, 163–165]

c) release of high energy water molecules in complex formation [2, 74]

d) release of strain energy in the macromolecular ring of the cyclodextrin [54, 55, 120, 125].

The Van der Waals interactions here include both permanent dipole-induced dipole interactions and London dispersion forces. Interactions between permanent dipoles of the guest and permanent dipoles of the host are not important, since guests without permanent dipoles can form stable complexes with cyclodextrins. The strength of both the permanent dipole-induced dipole interactions [166] and the

23

London dispersion forces [167] are proportional (approximately) to the reciprocal 6th power of the distance between the two components and to the polarizabilities of the two components. Thus, the magnitudes of these interactions can be large in inclusion complexes, since the internal diameter (4.5 Å for α-cyclodextrin, 7.0 Å for β-cyclodextrin) can lead to only a small distance between the guest and the wall of the cyclodextrin. For a series of substituted phenyl acetates, an approximately linear relationship was observed between the logarithm of the dissociation constant of the cyclodextrin-substrate complex and the molar refraction of the substrate. This fact indicates a major role of Van der Waals forces, since the molar refraction of a substrate is related to its polarizability [74]. Additionally, the dissociation constant of inclusion complexes for a series of *para*-substituted benzoic acids are correlated by Hammett substituent constants which may also be related to polarizability [89]. Thus, undoubtedly Van der Waals interactions play some role in inclusion complex formation. The importance of these forces has also been shown in other molecular complexes [168, 169].

Hydrogen bonding also functions in inclusion complex formation, since *t*-butyl alcohol, which hydrogen bonds less easily than *t*-butyl hydroperoxide, does not form an inclusion complex with cyclodextrin, whereas the latter does [170].

Although Van der Waals forces and hydrogen bonding obviously function, these are not enough to explain the fact that inclusion complexes are formed most readily in aqueous solution. No interactions between benzoic acid and cyclodextrins were observed in solvents such as chloroform, carbon tetrachloride, ether, dioxane, and benzene [151]. Furthermore, the dissociation constants of inclusion complexes are much larger in dimethyl sulfoxide than in water; K_d of the β-cyclodextrin-*m-t*-butylphenyl acetate complex and of the β-cyclodextrin-anisole complex at 25° are 18 mM and 400 mM in dimethyl sulfoxide but only 0.1 mM and 5 mM in water [70]. Thus water seems to play an important role in inclusion complex formation, which can be well interpreted in terms of proposals (c) and (d) above. Besides, substrates with apolar groups favor the inclusion process [75, 171], indicating apolar-like interactions.

In aqueous solution, several water molecules are accommodated in the cavity of cyclodextrin [172]. However, the structure of the water in the cavity can not be the same as that of bulk water, since the cyclodextrin cavity is apolar and of restricted size. Bender and coworkers suggested that water molecules in the cyclodextrin cavity are enthalpy rich, since they cannot form their full complement of hydrogen bonds to adjacent water molecules [2, 74]. Consequently, inclusion complex formation involves the replacement of these high enthalpy water molecules by guest compounds, resulting in a favorable enthalpy change.

On the other hand, Saenger and coworkers proposed that cyclodextrin can release the strain energy of its macrocyclic ring on inclusion complex formation [54, 55, 120, 125]. According to the results of their X-ray crystallography and potential energy calculations, the macrocyclic conformation of the α-cyclodextrin torus in aqueous solution before inclusion complex formation (assumed to be identical with that in α-cyclodextrin hexahydrate) is less symmetrical and of higher energy than the conformation of the α-cyclodextrin in inclusion complexes with iodine, 1-propanol, methanol, or potassium acetate (see p. 20).

Several attempts were made to determine the extent of the contribution of each of the four kinds of interaction in inclusion complex formation. Bergeron and co-workers recently evaluated the contribution of strain energy relief and the release of high energy water in inclusion complex formation [84, 173]. Methylation of the hydroxyl groups of cyclodextrin, which is supposed to affect largely the dissociation constant of the cyclodextrin-substrate complex, if release of strain energy is important, exhibited only a minor effect on the dissociation constant. Thus, this effect is not the main driving force [173]. Furthermore, they proposed that the fact that p-nitrophenol binds with cyclodextrin more tightly than phenol whereas p-nitrobenzoic acid binds more weakly than benzoic acid is unfavorable for both the "strain energy relief" mechanism and the "release of high energy water" mechanism. Neither mechanism satisfactorily explains the stronger binding of p-nitrophenolate ion than of p-nitrophenol with cyclodextrin. Besides, it indicates the significance of the orientation of the dipoles of the guest compounds that both p-nitrophenol and p-nitrophenolate ion bind with cyclodextrin via the nitro group first (see p. 12). They proposed that the dipole-induced dipole interactions are most important for inclusion complex formation of these guest compounds [84].

Calculations of the magnitudes of contributions of interaction energies for inclusion complex formation between α-cyclodextrin and sodium benzenesulfonate were attempted by Harata [127]. Magnitudes of Van der Waals interactions and electrostatic interactions between host and guest molecules, and the difference between the solvation energy of the complex and the sum of the solvation energies of the two components in water were evaluated. Results of these calculations, made on two possible structures of the complex (6a, 6b), are shown in Table 8. These calculations indicate several things:

a) the structure (6a), in which the apolar benzene group is included in the cyclodextrin cavity, is more stable than the structure (6b), in which the polar sulfonate group is included, by about 8.7 kcal/mole

Table 8. Calculated energies of complex formation between α-cyclodextrin and sodium benzenesulfonate [a, b]

Structure of complex	Interaction energies (kcal/mol)			
	Van der Waals interaction	Electrostatic interaction	Change of solvation energy	Total
6a	−4.59	−0.03	−19.15	−23.78
6b	−2.92	−0.06	−12.05	−15.04

[a]
[b] From Harata, K.: Bull. Chem. Soc. Japan 49, 2066 (1976).

6a   6b

b) the change of solvation energy plays a major part in interaction energies

c) Van der Waals interactions have considerable importance, while electrostatic interactions are negligibly small

However, the calculations are too oversimplified to exactly evaluate the driving force
of inclusion complex formation, since neither the difference in enthalpy of the water
inside and outside the cavity nor the strain energy in the cyclodextrin was taken into
consideration.

A more comprehensive calculation of the interaction forces for complex formation was recently made by Tabushi et al. [148c]. The magnitude of the Van der Waals
interaction energy, solvation energy of apolar solutes in water, hydrogen bond energy
of the water molecules in the cavity of α-cyclodextrin hexahydrate, and the conformational energy of the cyclodextrin were all calculated. In complex formation of methyl
orange with α-cyclodextrin, the Van der Waals stabilization energy in the α-cyclodex-
trin-guest compound is larger than in the α-cyclodextrin hexahydrate (which was
taken as the structure of α-cyclodextrin before inclusion complex formation) by 8.35
kcal/mole. The increase in entropy accompanied by breaking the water clusters around
the apolar guest (+32.9 e. u.) is also important for stabilization. Surprisingly, the calculation by Tabushi et al. showed that the conformational energy in the α-cyclodex-
trin-methyl orange complex is larger than in the α-cyclodextrin hexahydrate by 3.83
kcal/mole. This result is obviously in conflict with the proposal by Saenger et al [54,
55, 120, 125] that the conformational energy is the main driving force.

The importance of classical apolar binding was indicated by measuring the
enthalpy and entropy change in complex formation of 1-adamantanecarboxylate with
α-cyclodextrin. This substrate is too bulky to be completely included in the cavity of
α-cyclodextrin and probably sits on top of the cavity. Its complex formation with
α-cyclodextrin showed a quite favorable entropy change (ΔS = +10 e. u.) but
only a small favorable enthalpy change (ΔH = -1.2 kcal/mole). Thus, more than
70% of the stabilization of the complex comes from entropy, which is associated with
a transfer of the substrate from aqueous medium to more apolar medium (the cavity).
On the other hand, complex formation of 1-adamantanecarboxylate with β-cyclodextrin, which has a cavity large enough to accommodate this bulky substrate, exhibited
a large favorable enthalpy change (ΔH = -4.7 kcal/mole) and a small unfavorable
entropy change (ΔS = -1 e. u.). Consequently, it was proposed that most of the
stabilization energy of the inclusion complex comes from apolar binding. Other factors such as Van der Waals interactions and London dispersion forces are largely
cancelled by the accompanying entropy loss [242c].

Several attempts were made to improve the binding of cyclodextrins with substrates through capping one side of the cyclodextrin torus by apolar moieties [148a,
174]. For example, capped β-cyclodextrin (7) bound sodium 1-anilino-8-naphthalene-
sulfonate 24 times better than parent β-cyclodextrin [148a]. This finding is favorable
for the proposal of high energy water release as the driving force of inclusion complex formation, since capping increases the apolar nature of the cavity, and thus make
the water molecules in the cavity still more unstable. On the other hand, the strain
release mechanism as the driving force is not in accord with this result, since a change

of macrocyclic conformation on complex formation should be hindered by bridging of two hydroxyl groups in (7). Thus, the binding of (7) with substrate would have become worse, if release of macrocyclic strain energy were important for binding. Obviously, this is not the case.

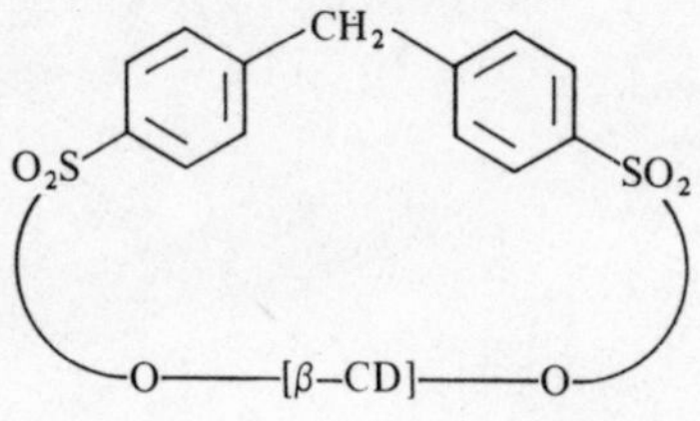

(7)

In conclusion, a variety of factors such as Van der Waals and London dispersion forces, release of high energy water, and possibly relief of strain energy are all operative. The relative contribution of each factor is still unknown. It is noteworthy that release of high energy water and relief of strain energy are associated with apolar binding and the lock and key mechanism in enzymatic reactions, respectively.

27

IV. Catalyses by Cyclodextrins Leading to Practical Usages of Cyclodextrins

It is well known that cyclodextrins can accelerate many kinds of reactions. The reactions accelerated by cyclodextrins range widely (see Table 9).

The first observation of an α-cyclodextrin-accelerated reaction was on the hydrolysis of ethyl p-chloromandelate [163, 164]. The addition of 1.32×10^{-3} M β-cyclodextrin accelerated its hydrolysis by 1.38 fold. Besides, after 50% conversion in the hydrolysis, the resulting acid and the remaining ester had a specific rotation of $[\alpha]_D^{25} =$ +0.38° and $[\alpha]_D^{25} = -0.17°$, respectively. This finding suggested that cyclodextrins can be an accelerating agent as well as asymmetric agent, although both the rate effect and optical yield were small.

The catalyses by cyclodextrins take place through inclusion complex formation of substrates with cyclodextrins discussed previously. This was shown by the following facts:

a) the reaction rate is not a linear function of cyclodextrin concentration, but it approaches a maximum value asymptotically with an increase of cyclodextrin concentration [53, 74, 108, 152, 176–186]

b) competitive inhibition was found by addition of organic compounds, which competitively bind in the cyclodextrin cavity [74, 184]

c) α-methyl glucoside, a monomolecular analog of cyclodextrins, exhibited no or much smaller effect on these reactions [74, 184]

Cyclodextrin-catalyzed reactions show many of the kinetic features shown by enzymatic reactions, including saturation [53, 70, 74, 107, 108, 152, 178, 179, 184 to 186], stereospecific catalyses [53, 74, 175, 176, 185, 186], and D.L-specificity [53, 107, 108, 163, 186] as well as substrate-catalyst complex formation, and competitive inhibition described above. Thus, cyclodextrin can serve as models of certain enzymes [2, 7].

Cyclodextrin-catalyzed reactions can be classified in the following two categories:

a) covalent catalyses, in which cyclodextrins catalyze reactions via formation of covalent intermediates

b) noncovalent catalyses, in which cyclodextrins provide their cavities as apolar or sterically restricted reaction fields without the formation of any covalent intermediates

Table 9. Reactions accelerated by cyclodextrins

Reactions	Substrates	Acceleration factor[a]	Kind of catalysis[b]	References
Cleavage of esters	Phenyl esters	300[c]	C	[52, 70, 74, 175–177]
	Mandelic acid esters	1.38[d]	U	[163]
Cleavage of amides	Penicillins	89[e]	C	[178, 179]
	N-Acylimidazoles	50[f]	C	[180]
	Acetanilides	16[g]	C	[181]
Cleavage of organophosphates	Pyrophosphates	>200[h]	C	[182–184]
	Methyl phosphonates	66.1[i]	C	[53, 185, 186]
Cleavage of carbonates	Aryl carbonates	7.45[j]	C	[185]
Cleavage of sulfates	Aryl sulfates	18.7[k]	N	[187]
Intramolecular acyl migration	2-Hydroxylmethyl-4-nitrophenyl trimethylacetate	6[l]	N	[188]
Decarboxylation	Cyanoacetate anions	44.2[m]	N	[189–191]
	α-Ketoacetate anions	3.95[n]	N	[189–191]
Oxidation	α-Hydroxyketones	3.3[o]	N	[192]

[a] Ratio of the rate catalyzed by cyclodextrin to the uncatalyzed reaction rate.
[b] C, N, and U, respectively, refer to covalent catalysis, noncovalent catalysis, and unknown.
[c] For m-nitrophenyl acetate hydrolysis by α-cyclodextrin in [74].
[d] For ethyl 4-chloromandelate hydrolysis by β-cyclodextrin.
[e] For 2-naphthylpenicillin hydrolysis by β-cyclodextrin in [76].
[f] For N-acetylimidazole hydrolysis by α-cyclodextrin.
[g] For 2,2.2-trifluoro-p-nitroacetanilide hydrolysis by α-cyclodextrin.
[h] For p-chlorophenyl pyrophosphate hydrolysis by β-cyclodextrin in [184].
[i] For bis(m-nitrophenyl)methylphosphonate hydrolysis by α-cyclodextrin.
[j] For bis(p-nitrophenyl)carbonate hydrolysis by α-cyclodextrin.
[k] For 2,4-dinitrophenyl sulfate hydrolysis by β-cyclodextrin.
[l] By α-cyclodextrin.
[m] For p-chlorophenylcyanoacetic acid decarboxylation by β-cyclodextrin in [191].
[n] For α-benzylacetoacetic acid decarboxylation by β-cyclodextrin in [190].
[o] For the oxidation of deoxyindole by β-cyclodextrin.

Cyclodextrins exhibit remarkable *ortho-para* selectivity in the chlorination of aromatic compounds by hypochlorous acid (HOCl) [193–195]. The catalyses by cyclodextrins in these reactions are obviously different than typical covalent catalyses. However, this can be classified as covalent catalysis rather than as noncovalent catalysis, since chlorination takes place via formation of a hypochlorite ester of cyclodextrin, a covalent intermediate. In the chlorination of anisole by hypochlorous acid, *para*-chlorination occurs almost exclusively in the presence of sufficient cyclodextrin, although in control experiments maltose had no effect on the product ratio (Table 10). This table shows that *ortho*-chlorination of anisole complexed with cyclodextrin is almost completely hindered, whereas *para*-chlorination of complexed anisole is 5.6 times faster than that for free anisole. Furthermore, the rate of chlorination of com-

Table 10. The effect of α-cyclodextrin on the *ortho-para* ratio in the chlorination of anisole by hypochlorous acid[a]

α-Cyclodextrin 10^{-3} M	Chloroanisole product ratio, *p:o*	% Anisole bound
0	1.48	0
0.933	3.43	20
1.686	5.49	33
2.80	7.42	43
4.68	11.3	56
6.56	15.4	64
9.39	21.6	72

[a] Reprinted with permission from Breslow, R., Campbell, P.: J. Amer. Chem. Soc. *91*, 3085 (1969). Copyright by the American Chemical Society.

plexed anisole is first-order in HOCl, although the rate of chlorination of free anisole in solution is second-order in HOCl. In the proposed mechanism, one of the secondary hydroxyl groups reacts with HOCl to form a hypochlorite group, which attacks the sterically favorable *para* position of the anisole molecule included in the cyclodextrin cavity in an intracomplex reaction. The participation of one of the secondary hydroxyl groups at the C-3 position in the catalysis was shown by the fact that dodecamethyl-α-cyclodextrin, in which all the primary hydroxyl groups and all the secondary hydroxyl groups at the C-2 positions are methylated, exhibited equal or larger *ortho-para* specificity than native α-cyclodextrin [195].

Interestingly, an O-alkylated polymer (8), produced by reacting α-cyclodextrin with epichlorohydrin under basic condition, exhibited better *ortho-para* specificity in chlorination than α-cyclodextrin itself [195]. A column of a resin of this polymer was loaded with anisole; then an aqueous solution of HOCl was passed through. The resulting chloroanisole was greater than 99% *para,* with less than 1% *ortho.* These columns can be used repeatedly with no sign of deterioration.

8

Furthermore, cyclodextrin enhances the hydrolysis of trichlorphon, which is probably attributable to the inclusion complex formation [196a]. (see Chap. VI for the mechanism).

β-Cyclodextrin catalyzes the electrophilic allylation of 2-methylhydronaphthoquinone with allyl or crotyl bromide in aqueous medium, giving the corresponding vitamin K_1 or K_2 analog in excellent yield. A solution of 5×10^{-2} M β-cyclodextrin, 2×10^{-1} M allyl bromide and 1×10^{-2} M 2-methyl-1,4-hydronaphthoquinone in a mixture of borate buffer (pH 9.0) and methanol (70v/30v) was stirred at room temperature for 9 hrs. under a nitrogen atmosphere in the dark. The products were 2-methyl-3-allylnaphthoquinone (40% yield) and 2-methylnaphthoquinone (54% yield). The allylation reaction in the absence of β-cyclodextrin, however, gave the K_1 or K_2 analog in a poor yield contaminated with considerable amounts of undersired by-products. A crotyl group is similarly introduced into the 3-position in the presence of β-cyclo-dextrin. The marked catalytic effect of the cyclodextrin was attributed both to the increase in nucleophilicity of the carbon atom on naphthohydroquinone monoanion in the cyclodextrin cavity and to the protection against oxidative cleavage of the included naphthoquinone derivatives [197].

In some reactions, however, cyclodextrins function as a retarder or inhibitor of reaction [51, 52, 67, 152, 198–205]. For example, the hydrolyses of ethyl amino-benzoates were almost completely inhibited by complexation with cyclodextrins [52, 67]. Quantitatively similar results were obtained for the hydrolyses of methyl benzoate, ethyl benzoate, ethyl *trans*-cinnamate, and methyl *m*-chlorobenzoate [51]. In addition, intramolecular carboxylate ion attack in glutaric acid esters [201], the decomposition of hydroperoxide [200], the benzidine rearrangement of hydrazo-benzene [202], the photochemical *cis-trans* isomerization of methyl orange [203], and the photodehalogenation of monohalobenzoic acids [204] are retarded or prohibited by cyclodextrins. The hydrolyses of aryl sulfates catalyzed by detergents such as N,N-dimethylhexadecylamine or N,N-dimethyl-N-hexadecyl-N-(4-imidaz-olium)methylammonium dichloride are also retarded by β-cyclodextrin, which was ascribed to competitive inclusion of the substrate and the detergent molecules in the β-cyclodextrin cavity [205]. Furthermore, the unsaturated fatty acids included in cyclodextrins are practically completely protected against oxidation even in pure oxygen [196b].

In addition to catalyses by cyclodextrins, complexes of cyclodextrins with drugs, insecticides, dyes and so on [206–229] exhibit many kinds of physicochemical and biochemical features, which are not shown in the absence of cyclodextrins. For example, the inclusion complex of a drug, N-(2,3-dimethylcyclohexyl)-N-methylanthra-nilic acid, with cyclodextrin, formed by heating the drug and the cyclodextrin at 60 °C, is highly soluble in water and thus is suitable for injections [224]. The clath-rates of resmethrin with cyclodextrin is about 1.5 times more effective in killing cockroaches than resmethrin itself [212]. Furthermore, β-cyclodextrin stabilizes co-enzyme A. Residual coenzyme A was 86.8–98.9% active after 1, 2, and 3 months at pH 5.0 in the presence of β-cyclodextrin, whereas it was 72.7–90.0% active in con-trol experiments without cyclodextrin [230]. These facts provide possibilities of important practical usage to cyclodextrins.

Furthermore, the increase of solubility of guest compounds in water by cyclo-dextrins makes it possible to carry out some reactions using these compounds in aqueous solution. For example, the fluorescent labeling of proteins and the plasma membrane by dansyl chloride can be done only by using a suspension of an organic solvent-water mixture, because of scant solubility of dansyl chloride in water. How-ever, the β-cyclodextrin-dansyl chloride complex can effectively label proteins and

31

plasma membranes in aqueous solution without using any organic solvent [231 to 233].

Another interesting use of cyclodextrin is as a ^{1}H-NMR shift reagent for hydrocarbons. Hydrocarbons are not in general amenable to use of lanthanide shift reagents or of aromatic solvents to bring about changes in NMR spectra useful in structure determination. The aromatic protons of p-cymene give rise to only a single band in a wide range of solvents. Lanthanide shift reagents are ineffective. However, addition of an excess amount of α-cyclodextrin exhibited a marked splitting of the aromatic spectrum as shown in Fig. 8. The chemical shift separation by cyclodextrin is highly temperature dependent, and is significantly greater at ambient temperature. Interestingly, β-cyclodextrin, which has a larger cavity than α-cyclodextrin, has almost no effect in chemical shift separation. The rigidity of the guest compound in the cyclodextrin cavity should be important. α-Cyclodextrin gives a well resolved spectrum of adamantane. However, chloroform, benzene, dioxane, and methanol gave a poorly resolved spectrum. Adamantane, a saturated, and approximately spherical hydrocarbon, can not be included in the cyclodextrin cavity owing to bulkiness. Probably, adamandane sits on the top of the cavity and takes a certain orientation [234].

There are several applications of cyclodextrins used in binding, especially in column chromatography described in Chap. III, 1.

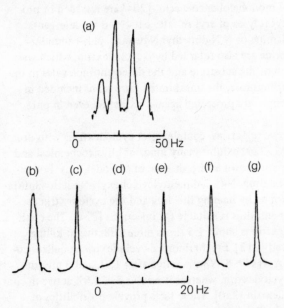

Fig. 8. 100 MHz ^{1}H-NMR spectra of the aromatic protons of p-cymene; (a) in the presence of α-cyclodextrin in D_2O at 50 °C, (b) as neat liquid, (c) in $CDCl_3$, (d) in $CDCl_3$/Eu(fod)$_3$, (e) in hexamethylphosphoramide, (f) in tetramethyl urea and (g) in methanol. From MacNicol, D. D.: Tetrahedron Let. *1975*, 3325

V. Covalent Catalyses

Covalent catalyses by cyclodextrin (C) proceed as shown in Scheme 1, where S and $S \cdot C$ represent the substrate and the (noncovalent) inclusion complex, respectively; $C - P_1$ is the covalent intermediate, which is an acyl-cyclodextrin in the hydrolyses of esters and amides and is cyclodextrin phosphate in the hydrolyses of organophosphates; P_1 and P_2 are the final products; and k_1, k_{-1}, k_2, k_3, and k_{un} are rate constants for the various processes indicated.

$$S + C \underset{k_{-1}}{\overset{k_1}{\rightleftharpoons}} S \cdot C \xrightarrow{k_2} \underset{+ P_2}{C^{'} - P_1} \xrightarrow{k_3} C + P_1$$

$$\downarrow k_{un}$$

$$P_1 + P_2$$

Scheme 1

Thus, the first step of covalent catalyses by cyclodextrins is complex formation between cyclodextrin and substrate. The second step is the nucleophilic attack by one of the hydroxyl groups of the cyclodextrins on the substrate resulting in a covalent intermediate $(C - P_1)$ and P_2. Then, the covalent intermediate $(C - P_1)$ hydrolyzed to the final product P_1, and C is regenerated.

The formation of the $S \cdot C$ complex preceding the catalytic reactions has been shown by saturation phenomena. The observed pseudo first-order rate constant, k_{obs}, asymptotically approaches a maximum value as the added (excess) cyclodextrin concentration increases. The formation of $C - P_1$ has been proved both by spectroscopic detection of cyclodextrin-benzoate and cyclodextrin-*trans*-cinnamate in the cyclodextrin-accelerated hydrolyses of phenyl benzoates [51] and *trans*-cinnamoylimidazole [180] and by isolation of acyl cyclodextrins in situ [51, 107, 108, 235].

Scheme 1 is identical with that in many enzymatic reactions. The rate constant, k_2, and the dissociation constant of the $S \cdot C$ complex, K_d, which is equal to k_1/k_{-1}, can be determined by the use of Lineweaver-Burk-type plots (Eq. 2) [236] under conditions that $[C]_0 \gg [S]_0$, where $[C]_0$ and $[S]_0$ are the initial concentrations of cyclo-

$$\frac{1}{k_{obs} - k_{un}} = \frac{K_d}{(k_2 - k_{un})[C]_0} + \frac{1}{k_2 - k_{un}} \tag{2}$$

33

dextrin and substrate, respectively. Plots of $1/(k_{obs} - k_{un})$ versus $1/[C]_0$ give values of k_2 and K_d from the intercept and slope. Alternatively, Eadie-type plots corresponding to equation (3) [237] can be used to reduce statistical errors in the values of k_2 and K_d.

$$k_{obs} - k_{un} = - \frac{K_d(k_{obs} - k_{un})}{[C]_0} + (k_2 - k_{un}) \tag{3}$$

When $[S]_0 \gg [C]_0$, $[C]_0$ in equation (2) and (3) is simply replaced by $[S]_0$.

One of the most important points in covalent catalyses by cyclodextrins is that the reactivity of the substrate can be almost fully explained in terms of the geometry of the complex, e.g. the orientation of the reacting center of the substrate with respect to the catalytically active secondary hydroxyl groups of the cyclodextrin in the complex. Information on the geometry of the complex, which is much more definite than the geometry of reactant and catalyst in solution, can be easily obtained by the use of molecular models.

1. Hydrolyses of Phenyl Esters

Hydrolyses of phenyl esters are one of the most precisely and extensively studied systems of many cyclodextrin-accelerated reactions [51, 74, 175—177]. Much information was obtained from these systems, most of which is applicable to other covalent catalyses by cyclodextrins.

First of all, the first step of cyclodextrin-catalyzed hydrolyses of phenyl esters, the cleavages of these esters resulting in acyl-cyclodextrins, is described. Table 11 lists

Table 11. Values of k_{un}, k_2, and K_d in the α-cyclodextrin-accelerated cleavages of phenyl acetates[a,b]

Acetate	k_2 (10^{-2} sec^{-1})	k_{un} (10^{-4} sec^{-1})	$\dfrac{k_2}{k_{un}}$	K_d (10^{-2} M)
Phenyl	2.19	8.04	27	2.2
m-Tolyl	6.58	6.96	95	1.7
p-Tolyl	0.22	6.64	3.3	1.1
m-t-Butylphenyl	12.9	4.90	260	0.2
p-t-Butylphenyl	0.067	6.07	1.1	0.65
m-Nitrophenyl	42.5	46.4	300	1.9
p-Nitrophenyl	2.43	69.4	3.4	1.2
m-Carboxyphenyl	5.55	8.15	68	10.5
p-Carboxyphenyl	0.67	12.5	5.3	15.0

[a] Reprinted with permission from VanEtten, R. L., Sebastian, J. F., Clowes, G. A., Bender, M. L.: J. Amer. Chem. Soc. 89, 3242 (1967). Copyright by the American Chemical Society.

[b] pH 10.6, 25 °C, I = 0.2 M.

the values of k_{un}, k_2, and K_d, which were determined by Lineweaver-Burk type plots corresponding to equation 2. Both α- and β-cyclodextrins accelerate the cleavage of a variety of phenyl esters. The acceleration factors (k_2/k_{un}) vary widely from 300 fold for m-nitrophenyl acetate to 1.1 fold for p-t-butylphenyl acetate. The K_d values for β-cyclodextrin complexes are in general 2–10 times smaller than those for α-cyclodextrin complexes.

In complex formation of the substrate with cyclodextrin, the phenyl portion of the substrate is included in the cyclodextrin cavity from the wide secondary hydroxyl group side. Then, the catalyses by cyclodextrin (cleavages of the phenyl esters) proceed through nucleophilic attack by a secondary hydroxyl group (in the ionic state) on the carbonyl carbon atom of substrate [51]. This was shown by

a) the process is governed by a functional group of pK_a 12.1

b) heptamesyl-β-cyclodextrin [238], in which all primary hydroxyl groups are blocked, causes as large an acceleration as native β-cyclodextrin

c) dodecamethyl-α-cyclodextrin [239], in which all primary hydroxyl and all secondary hydroxyl groups at the 3 position are blocked, cause a small inhibition of cleavage rather than an acceleration

The comparatively low value of the pK_a of the secondary hydroxyl groups on the cyclodextrin is due to stabilization of the alkoxide ion by means of intramolecular hydrogen bonds to neighboring hydroxyl groups, which is characteristic of cyclodextrin (see Chap. II) as well as to the combined inductive effects of the electronegative oxygen atoms. This pK_a (12.1) is consistent with that (12.2) found by Chin et al. [52]. As mentioned before, catalyses by secondary hydroxyl groups involve nucleophilic attack on the carbonyl carbon atom. This is certainly so since there is no kinetically important D_2O solvent isotope effect [240]. The rate constant for the hydrolysis of m-t-butylphenyl acetate in H_2O was larger than in D_2O by 3.2 fold, which can be wholly accounted for by the difference between the pK_a of cyclodextrin in H_2O and that in D_2O [241].

Fig. 9 shows the plot of the acceleration by 10^{-2} M cyclodextrin versus the Hammett substituent constant [74]. Obviously, acceleration does not follow the Hammett relationship. Interestingly, however, a fair Hammett relationship was observed for the rates of hydrolyses of these phenyl esters in the absence of cyclodextrin. Likewise, the rates of hydrolyses in the presence of α-methyl glucoside, a monomolecular analog of cyclodextrins but without a cavity, followed the Hammett relationship. Thus, the acceleration by cyclodextrins can not be ascribed to electronic effects.

The results in Table 11 and Fig. 9 can be summarized as follows:

a) *meta*-substituted phenyl esters show larger acceleration by cyclodextrin than corresponding *para*-compounds

b) *meta/para* specificity is larger for esters with more bulky substituents

c) the reactivity of unsubstituted esters is between that of the *meta*- and *para*-substituted compounds

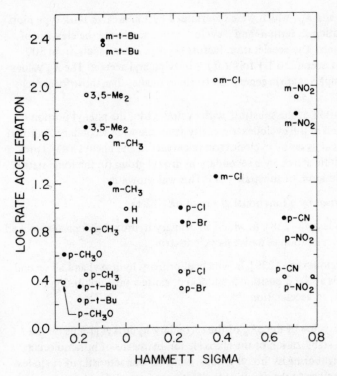

Fig. 9. Rate accelerations by 1% α-cyclodextrin, ○, and 1% β-cyclodextrin, ●, in the liberation of phenols from substituted phenyl acetates at pH 10.6, 25°. Reprinted with permission from VanEtten, R. L., Sebastian, J. F., Clowes, G. A., Bender, M. L.: J. Amer. Chem. Soc. *89,* 3242 (1967). Copyright by the American Chemical Society

All these *meta/para* specificities are due to the geometry of the inclusion complexes, especially the distance between the electrophilic center (the carbonyl carbon atom in phenyl esters) and the nucleophilic catalytic center (a secondary hydroxyl group of the cyclodextrin). Fig. 10 shows Corey-Pauling-Koltun molecular models of these complexes. The ester function of the *meta*-substituted compound (large acceleration) is located very close to the secondary hydroxyl groups in the inclusion complex, whereas the ester function of the *para*-substituted compound (small acceleration) is located at a considerable distance from the secondary hydroxyl groups. These results of molecular model studies were supported by X-ray crystallography of the *p*-iodo-aniline complex [131]. The distance between the amino group and the secondary hydroxyl groups in the α-cyclodextrin-*p*-iodoaniline complex, in which the iodo-benzene group is situated in the cavity with the amino group sticking out of the cavity, is larger than 4.0 Å, which would make catalysis of phenyl ester hydrolysis impossible.

The importance of orientation of the catalytic site (the secondary hydroxyl group) around the reactive site (the carbonyl portion of a phenyl ester) in the cleavages of phenyl esters was also indicated by comparing the effects of *p*-carboxyphenyl esters. β-Cyclodextrin showed 5.3 fold *acceleration* in the cleavage of *p*-carboxyphenyl ace-

Fig. 10. Corey-Pauling-Koltun models of the α-cyclodextrin-p-t-butylphenyl acetate complex (left) and the α-cyclodextrin-m-t-butylphenyl acetate complex (right). Reprinted with permission from VanEtten, R. L., Sebastian, J. F., Clowes, G. A., Bender, M. L.: J. Amer. Chem. Soc. *89*, 3242 (1967). Copyright by the American Chemical Society

tate. However, in the cleavages of *p*-carboxyphenyl 2-methylpropionate and *p*-carboxyphenyl 3,3-dimethylbutyrate, which have more hydrophobic ester functions than the acetate has, it exhibited retardation (k_2/k_{un} = 0.68 and 0.19, respectively). Probably the hydrophobic ester functions of these esters are included in the cyclodextrin cavity, since hydrophilic carboxylate ion would not be readily included in the cavity because of its solvation requirement. In these conformations of the inclusion complexes, the carbonyl carbon of the ester (the electrophile) is too far from the secondary hydroxyl group of the cyclodextrin (the nucleophile) for effective catalysis to take place. Thus, these can be called nonproductive complexes [74].

Acceleration by cyclodextrin in the cleavage of phenyl esters is thus attributed to the proximity effect between the catalytic and reactive sites. This conclusion is confirmed by Table 12. Conversion of intermolecular catalysis to intracomplex catalysis is accomplished by multiplying by 10 M, which is the averaged ratio of an intramolecular catalyst to an intermolecular catalyst. This calculation gave a value close to the experimental value of k_2(lim), where k_2(lim) is the limiting k_2 value corresponding to the complete ionization of the hydroxyl groups of the cyclodextrin [242a].

The origin of acceleration of the cleavage of phenyl esters by cyclodextrins was examined more carefully. The activation enthalpy ($\Delta H^{\ddagger}$) and entropy ($\Delta S^{\ddagger}$) for the α-cyclodextrin-accelerated cleavage of several (*p*-methyl, *m*-methyl, *p*-nitro, *m*-nitro,

Table 12. Kinetic factors responsible for the difference in rate of liberation of *m-t*-butylphenol from *m-t*-butylphenyl acetate by hydroxide ion and α-cyclodextrin[a]

Rate constant of hydroxide ion catalysis	1.2 M^{-1} sec^{-1}
Conversion to rate constant of alkoxide ion reaction of pK_a 12.1 (fourfold) from a hydroxide ion reaction (pK_a 15.7)	4.8 M^{-1} sec^{-1}
Conversion to an intramolecular reaction from an intermolecular reaction (10 M)	48 sec^{-1}
Experimental α-cyclodextrin, k_2(lim)	13 sec^{-1}

[a] From Bender, M. L.: "Mechanisms of Homogeneous Catalysis from Protons to Proteins". New York: Wiley-Interscience 1971.

37

and *m*-chloro) phenyl acetates as well as for the alkaline hydrolyses were determined. Fig. 11 depicts the relationship between the acceleration by α-cyclodextrin at 25 °C (k_2/k_{un}) and the difference in the activation parameters between the α-cyclodextrin reactions and the alkaline hydrolyses ($\Delta\Delta H^{\ddagger}$ and $\Delta\Delta S^{\ddagger}$). Comparison of $\Delta\Delta H^{\ddagger}$ and $\Delta\Delta S^{\ddagger}$ instead of $\Delta H^{\ddagger}$ and $\Delta S^{\ddagger}$ was made here for normalization of the values with respect to different leaving groups. Clearly, $\Delta\Delta H^{\ddagger}$ decreases (a change favorable for reaction) linearly with the logarithm of k_2/k_{un}. The *meta*-compounds, which exhibit large acceleration by α-cyclodextrin, have small values of $\Delta\Delta H^{\ddagger}$, whereas the *para*-compounds, which exhibit small acceleration, have large values of $\Delta\Delta H^{\ddagger}$. The activation entropy term ($-T\Delta\Delta S^{\ddagger}$), however, linearly increases (unfavorable for reaction) with the logarithm of k_2/k_{un}. Thus, whereas the activation enthalpy governs the stereospecific acceleration of the cleavage of phenyl esters by cyclodextrin, the activation entropy term partly compensates the enthalpy term. Cyclodextrins are thus good enzyme models, especially for the examination of the effect of the formation of the enzyme-substrate complex on enzymatic acceleration [242b].

Cyclodextrin-accelerated cleavages of phenyl esters (acyl transfer from an ester to cyclodextrin) take place also in dimethyl sulfoxide (DMSO) $-H_2O$ solutions [70]. In 60% DMSO, the rate constant of the β-cyclodextrin-accelerated cleavage of *m-t*-butylphenyl acetate (k_2) is 3.8×10^{-1} sec^{-1}, which is about 500 times that of the cleavage in the absence of β-cyclodextrin ($k_{un} = 7.5 \times 10^{-4}$ sec^{-1}). (This is approximately twice that seen in aqueous solution.) Here, sodium carbonate and sodium

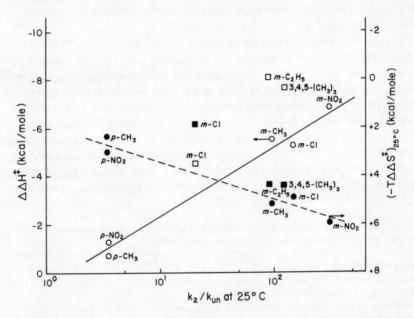

Fig. 11. Relations between the acceleration by cyclodextrins and the activation terms in the cyclodextrin-accelerated cleavages of phenyl acetates at 25 °C: $\Delta\Delta H^{\ddagger}$ and $\Delta\Delta S^{\ddagger}$ are the difference between the values for the cyclodextrin reactions and those for the corresponding alkaline hydrolyses; the round points refer to α-cyclodextrin and the square points refer to β-cyclodextrin. From Komiyama, M., Bender, M. L.: J. Amer. Chem. Soc., Submitted for publication (1977)

Table 13. Rate constants for the α-cyclodextrin-catalyzed hydrolysis of phenyl benzoates[a, b]

Benzoate	k_{un} $(10^{-4} \sec^{-1})$	$k_{acylation}$[c] $(10^{-4} \sec^{-1})$	$k_{deacylation}$[d] $(10^{-4} \sec^{-1})$
m-Nitrophenyl	15.4	1,400	4.6
m-Chlorophenyl	5.5	390	4.6
m-t-Butylphenyl	1.2	140	4.4

[a] From Van Etten, R. L., Clowes, G. A., Sebastian, J. F., Bender, M. L.: J. Amer. Chem. Soc. *89*, 3253 (1967).
[b] pH 10.6, 25°, I = 0.2 M.
[c] The rate constant for the appearance of phenolate ion in the presence of 0.01 M α-cyclodextrin.
[d] The rate constant for the appearance of benzoate ion.

borate buffers were used, corresponding to an aqueous pH of 9.5. The value of k_2 showed a bell-shaped curve as a function of solvent composition. k_2 increased from $8 \times 10^{-3} \sec^{-1}$ in H_2O to a maximum of $3.8 \times 10^{-1} \sec^{-1}$ in 60% DMSO, but after that the rate constant decreased. k_{un} also changed from $3 \times 10^{-5} \sec^{-1}$ in H_2O through $7.5 \times 10^{-4} \sec^{-1}$ in 60% DMSO to a maximum $1.3 \times 10^{-3} \sec^{-1}$ in 80% DMSO before decreasing at higher DMSO concentration.

The first step in the cyclodextrin-catalyzed hydrolyses of phenyl esters, the cleavages of phenyl esters, producing acyl-cyclodextrins, is followed by hydrolysis of this intermediate to acylate ion and regenerated cyclodextrin. In the cyclodextrin-catalyzed hydrolysis of phenyl benzoates, fast release of phenolate ion was followed by much slower release of benzoate ion. Interestingly, the rate of the second step is independent of substituents in the phenyl portion of the molecules, although the rate of the first step is largely affected by such substitution (Table 13). This indicates formation of a common intermediate, cyclodextrin-benzoate [51]. The same argument is used in enzymatic reactions. Furthermore, cyclodextrin-benzoate [51], cyclodextrin-*trans*-cinnamate and cyclodextrin-acetate [235] were synthesized by reactions of cyclodextrin with the corresponding phenyl esters.

Table 14. D_2O solvent isotope effects in the hydrolysis of β-cyclodextrin-*trans*-cinnamate as a function of pH[a, b]

pH	$k_{obs}(H_2O)/k_{obs}(D_2O)$	pH	$k_{obs}(H_2O)/k_{obs}(D_2O)$
10.0	5.2	12.5	2.9
10.5	5.2	12.7	2.5
11.2	5.2	12.9	2.3
12.2	4.8	13.2	2.1
12.4	3.8	13.4	2.1

[a] From Komiyama, M., Bender, M. L.: Bioorg. Chem. *6*, 323 (1977).
[b] 25 °C, I = 0.2 M.

The hydrolyses of acyl-cyclodextrins are also governed by a functional group of pK_a 12.1, which is identical with that of the secondary hydroxyl groups [51]. Table 14 shows the D_2O solvent isotope effect in the hydrolysis of β-cyclodextrin-*trans*-cinnamate [180]. Below pH (pD) 11.2, $k_{obs}(H_2O)/k_{obs}(D_2O)$ remains constant at 5.2. Above pH (pD) 11.2, this ratio gradually decreases with pH (pD) to a limiting value of 2.1. This fact shows that the hydrolysis of the acyl-cyclodextrin involves participation of an alkoxide ion as intramolecular general-base catalyst [243]. Alkaline hydrolysis, which is prevented by electrostatic repulsion on ionization of hydroxyl groups, can take place simultaneously with general base catalysis.

As described above, cyclodextrin-catalyzed hydrolysis proceeds quite similarly to enzymatic hydrolysis. Thus, it is interesting to compare the cyclodextrin-catalyzed rate constants with the rate constants for the corresponding alkaline-catalyzed hydrolysis of the same substrate, as shown in Table 15. Here, the second-order rate constants for the cyclodextrin reactions were obtained by division of $k_2(\lim)$ by the dissociation constant, K_d. The cyclodextrin reactions were 10^3- to 10^4-fold superior to hydroxide ion reactions, while the chymotrypsin reactions are 10^4- to 10^6-fold superior to hydroxide ion reactions. Thus, with respect to the rate enhancement, cyclodextrins seem to be as effective as chymotrypsin. However, there are two important differences between the cyclodextrin-catalyzed hydrolysis and the chymotrypsin-catalyzed hydrolysis:

a) The rate constant for the cyclodextrin reaction was determined at pH 13, its maximum, while that for the chymotrypsin reaction was determined at pH 8, its maximum

b) The rate constant k_3 is much smaller than k_2 in the cyclodextrin reactions, resulting in its inefficiency as a catalyst

Considering the fact that the hydrolysis of the acyl-cyclodextrin is slower than the hydrolysis of the parent phenyl ester, cyclodextrin can not be considered to be a true catalyst of hydrolysis.

Table 15. Second-order rate constants for the reactions of substrates with cyclodextrins and with chymotrypsin[a]

Substrate	Accelerating agent	Catalyzed rate/ hydroxide ion rate
m-t-Butylphenyl acetate	α-Cyclodextrin	1.8×10^3
m-t-Butylphenyl acetate	β-Cyclodextrin	2.6×10^4
Acetyltryptophanamide	Chymotrypsin	4×10^4
Acetyltryptophan ethyl ester	Chymotrypsin	1×10^6
Acetyltyrosine ethyl ester	Chymotrypsin	3×10^4

[a]　From VanEtten, R. L., Clowes, G. A., Sebastian, J. F., Bender, M. L.: J. Amer. Chem. Soc. 89, 3253 (1967).

Many attempts have been made to improve the efficiency of cyclodextrin as a catalyst or an enzyme model by overcoming these two defects. These attempts involved the attachment of other functional groups covalently or noncovalently to cyclodextrins. Those that were successful will be described in Chap. VIII.

2. Hydrolyses of Amides

Cyclodextrins accelerate the hydrolyses of the strained β-lactam ring of penicillins (9) [178, 179], the cleavage of N-acylimidazoles [180], and the hydrolyses of acetanilides [181].

9   10

The hydrolysis of (9) to penicilloic acid (10) is catalyzed by cyclodextrins. Cyclodextrins can therefore be considered a model of the enzyme penicillinase (β-lactamase). Table 16 shows the catalytic rate constant (k_2), acceleration (k_2/k_{un}), and the dissociation constant (K_d) for the β-cyclodextrin-catalyzed cleavage of (9). β-Cyclodextrin accelerates β-lactam cleavage (20–90 fold) compared with alkaline hydrolysis. Variation of the penicillin side chain (R) produces a larger effect on K_d than on k_2/k_{un}. The orientation of the β-lactam ring in the inclusion complex is affected minimally by the stereochemical requirements of binding, since the flexible molecular structure of (9) allows the carbonyl carbon atom access to the catalytic hydroxyl groups of cyclodextrin.

The presence of a covalent intermediate (penicilloyl-β-cyclodextrin) in the reaction pathway was demonstrated by:

a) a faster loss of (9) than the formation of final product (10)

b) the existence of an initial induction period in the rate of formation of (10) in the presence of an initial excess of (9)

c) an assay which is specific for penicilloyl derivatives (amides and esters)

As shown in Fig. 12, the concentration of the covalent intermediate gradually decreases after passing through a maximum, when the initial concentration of the substrate is larger than that of cyclodextrin. This fact indicates effective regeneration of cyclodextrin by hydrolysis of the covalent intermediate. Thus, cyclodextrin in this case can be called a true catalyst.

41

Table 16. Catalytic rate constants and accelerations in the β-cyclodextrin-catalyzed hydrolyses of penicillins[a]

Penicillins	Side chain, R	k_2 (min^{-1})	k_2/k_{un}	K_d (10^{-3} M)
Methylpenicillin	CH_3-	0.151	37	33
n-Pentylpenicillin	$CH_3(CH_2)_4-$	0.230	66	41
n-Nonylpenicillin	$CH_3(CH_2)_8-$	0.179	47	21
Benzylpenicillin	$C_6H_5CH_2-$	0.324	77	43
Diphenylmethylpenicillin	$(C_6H_5)_2CH-$	0.161	34	4.7
Triphenylmethylpenicillin	$(C_6H_5)_3C-$	0.624	40	3.85
Methicillin		0.074	21	12.8
1-Naphthylpenicillin		0.109	31	16
2-Naphthylpenicillin		0.427	89	75
4-Phenylphenylpenicillin		0.263	54	38
Ancillin		0.200	63	13.2

[a] From Tutt, D. E., Schwartz, M. A.: J. Amer. Chem. Soc. 93, 767 (1971).
[b] pH 10.24, 31.5°, I = 1.0 M.

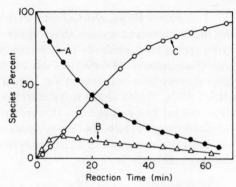

Fig. 12. Variation of the concentration of benzylpenicillin (curve A, black circles) and benzyl-penicilloic acid (curve C, open circles), expressed as per cent initial penicillin concentration, as a function of time in the presence of an initial excess of penicillin. Curve B corresponds to the variation of intermediate (penicilloyl-β-cyclodextrin) concentration, expressed as percent initial penicillin concentration, as a function of time. Reprinted with permission from Tutt, D. E., Schwartz, M. A.: J. Amer. Chem. Soc. *93*, 767 (1971). Copyright by the American Chemical Society

The N-acylimidazoles (formed as intermediates [244] of phenyl ester hydrolyses in nucleophilic catalysis by imidazole [244, 245]) are hydrolyzed through acid-catalyzed, neutral, and base-catalyzed mechanisms [246, 247]. Both α- and β-cyclodextrins accelerate the cleavages of amide bonds in N-*trans*-cinnamoylimidazole and N-acetylimidazole (Table 17) [180]. Covalent catalyses by cyclodextrin were definitely confirmed by spectrophotometric methods in these reactions. The absorbance at 335 nm rapidly decreased with time in the β-cyclodextrin-catalyzed hydrolyses of N-*trans*-cinnamoylimidazole, which corresponds to the disappearance of the substrate. After 1 min, the absorption spectrum of the solution was identical with that of an authentic sample of β-cyclodextrin-*trans*-cinnamate, synthesized in situ, plus that of imidazole. Then, the absorbance at 300 nm decreased much more slowly than in the first step, the rate of which was identical with that of the hydrolysis of an authentic sample of β-cyclodextrin *trans*-cinnamate. The final spectrum of the solution was identical with that of *trans*-cinnamate ion plus that of imidazole.

Table 17. Catalytic rate constants and accelerations in the cyclodextrin-catalyzed cleavage of N-acylimidazoles [a, b]

Substrate	Cyclodextrin	k_2 $(10^{-2} \text{sec}^{-1})$	k_2/k_{un}	K_d (10^{-2}M)
N-*trans*-cinnamoylimidazole[c]	α-Cyclodextrin	2.8	28	4.3
	β-Cyclodextrin	3.8	38	1.4
N-acetylimidazole[d]	α-Cyclodextrin	0.65	50	34
	β-Cyclodextrin	0.36	28	14

[a]   From Komiyama, M., Bender, M. L.: Bioorg. Chem. *6*, 323 (1977).
[b]   25°, I = 0.2 M, all rate constants are extrapolated to zero buffer concentration.
[c]   pH 9.0.   [d]   pH 7.0.

As described above, cyclodextrins accelerate the cleavages of the strained β-lactan ring of penicillins and acylimidazoles. However, the first example of the catalyses of hydrolyses of usual amides by cyclodextrin was the cyclodextrin-catalyzed hydrolyses of acetanilides. As shown in Table 18, the hydrolysis of p-nitrotrifluoroacetanilide catalyzed by α-cyclodextrin was 16 fold faster than its alkaline hydrolysis at pH 6.0, 30 °C. Hydrolyses of trifluoroacetanilide and m-nitrotrifluoroacetanilide were also catalyzed by α-cyclodextrin. However, cleavage of p-nitroacetanilide, a less activated substrate, was retarded by α-cyclodextrin (Table 18) [181].

The α-cyclodextrin-catalyzed hydrolysis of p-nitrotrifluoroacetanilide proceeds as shown in Scheme 2:

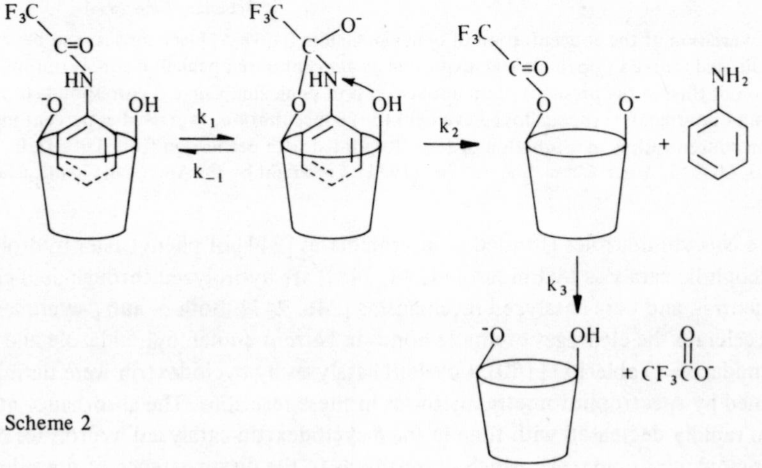

Scheme 2

Table 18. Values of k_2, k_{un}, and K_d for the cleavages of acetanilides[a]

Substrate	Temp. °C	pH (pD)	k_2 (or k_{obs}) (10^{-5} sec^{-1})	k_{un} (10^{-5} sec^{-1})	k_2/k_{un}	K_d (10^{-2} M)
p-Nitrotrifluoro-acetanilide	30	6.0	11	0.62	16	6.2
	30[b]	6.0[b]	2.6[b]	0.13[b]	20[b]	6.1[b]
	70	6.0	430	46	9.3	14
p-Nitroacetanilide	70	12.3	130[c]	165	–	–
	70	12.3	95[d]	165	–	–
Trifluoroacetanilide	70	9.0	41[c]	28	–	–
	70	9.0	59[d]	28	–	–
m-Nitrotrifluoro-acetanilide	70	9.0	72[c]	29	–	–
	70	9.0	95[d]	29	–	–

[a] From Komiyama, M., Bender, M. L.: J. Amer. Chem. Soc. *99*, 8021 (1977).
[b] In D_2O.
[c] Values obtained by extrapolating k_{obs} in the presence of 0.03 M α-cyclodextrin to zero buffer concentration.
[d] Values of k_{obs} extrapolated to zero buffer concentration in the presence of 0.06 M α-cyclodextrin.

Nucleophilic attack by a secondary hydroxyl ion at the carbonyl carbon atom of the substrate included in the cavity of α-cyclodextrin results in a tetrahedral intermediate, the breakdown of which to the acyl-α-cyclodextrin and aniline proceeds through general acid catalysis by the unionized secondary hydroxyl groups. The D_2O effect is in accord with the general acid catalysis in the breakdown of the tetrahedral intermediate. The D_2O effect for the α-cyclodextrin-accelerated cleavage of p-nitro-trifluoroacetanilide (the acylation step of the α-cyclodextrin-catalyzed hydrolysis) (4.2) is considerably larger than that for the β-cyclodextrin-accelerated cleavage of trans-cinnamoylimidazole (3.4) [180] and that for the α-cyclodextrin-accelerated cleavage of m-t-butylphenyl acetate (3.2) [240]. The general acid catalysis by the unionized secondary hydroxyl groups was required in the cyclodextrin-catalyzed cleavage of acetanilides because of their poor leaving groups. However, no general acid catalyses were observed in the cyclodextrin-accelerated cleavages of acylimidazoles and phenyl esters, since they have good leaving groups. The difference of the catalytic rate constant for the cleavages of these compounds in H_2O and in D_2O is due to difference of the pK_a of the hydroxyl groups of cyclodextrins in H_2O and in D_2O.

The kinetically determined dissociation constant of the complex between α-cyclodextrin and p-nitrotrifluoroacetanilide (K_d) was constant at 6.2×10^{-2} M irrespective of pH, which was identical with the K_d determined by the spectrophotometric method (6.6×10^{-2} M) within experimental error. The anionic species of p-nitrotrifluoroacetanilide

binds with α-cyclodextrin more strongly than the neutral species by about 4 fold ($K_d = 1.7 \times 10^{-2}$ M).

In the hydrolysis of p-nitrotrifluoroacetanilide, α-cyclodextrin operates as a true catalyst. The result of an experiment, where a substrate concentration larger than the cyclodextrin concentration was employed, showed that the concentration of α-cyclodextrin remains the same as the initial concentration during the process. This means that α-cyclodextrin is indeed regenerated by the rapid hydrolysis of the acyl-α-cyclodextrin. Thus, the rate-determining step is the acylation step. In the cyclodextrin-accelerated hydrolyses of phenyl esters, however, the deacylation step is rate-determining. The process involves esterification of the catalyst by the amide substrate, followed by the hydrolysis of the esterified catalyst, which is consistent with serine protease-catalyzed hydrolyses of amides. These findings indicate that cyclodextrins are indeed excellent models of hydrolytic enzymes since with both catalysts, esters show rate-determining deacylation, whereas amides exhibit rate-determining acylation [181].

3. Hydrolyses of Organophosphates and Carbonates

Diaryl pyrophosphates (11) are very stable in neutral or alkaline solution. Divalent metal ions accelerate their hydrolyses to a great extent. However, cyclodextrins exhibit still larger accelerations [182—184] (Table 19).

Reaction proceeds as indicated in Scheme 3, where an unsymmetrical pyrophosphate is shown as substrate to avoid confusion. Complex formation between the cyclodextrin and the substrate (11) is followed by nucleophilic attack by a hydroxyl group of cyclodextrin, resulting in the formation of cyclodextrin monophosphate (12) with release of phosphate ion (13). Then, (12) forms, with loss of phenol, a cyclic phosphate (14), which is immediately hydrolyzed into a simple phosphate ester (15). The intramolecular displacement within the covalent intermediate (12) is similar to the formation of cyclic phosphates from nucleotide phosphate diesters.

There is no doubt of the participation of hydroxyl group(s) of cyclodextrins in the first slow step of the cyclodextrin-accelerated cleavages of pyrophosphates. However, two other factors may contribute to this process;

a) Since the reaction is carried out at pH 12.0, which is almost identical with the pK_a of cyclodextrin, there are both hydroxyls and alkoxide ions in solution. Thus, the pyrophosphate, rigidly held between the $-OH$ and the $-O^-$ in the cyclodextrin cavity, is ruptured by a bifunctional catalytic mechanism, where the hydroxyl group and the alkoxide ion function as acid and base catalysts.

b) Calcium ion is bound to the pyrophosphate and labilizes the P—O—P bond by suppressing the $^-O-P=O$ resonance (general metal ion catalysis). Naturally, the Ca^{2+} is also surrounded by a water shell of first and higher order co-ordination, which should be several Å thick. On the other hand, the formation of the inclusion complex (= nonaqueous solvation) requires a partial removal of the water coordinated to Ca^{2+}, the water being removed at one side of the ion only. This, in turn, must tend to shift the Ca^{2+} ion in the direction from the cyclodextrin towards the water. The ion, therefore, will no longer be in a symmetrical position with respect to the pyrophosphate. Since the coulomb forces in the immediate neighbourhood of the ion are enormous,

Table 19. Accelerations of the cleavage of pyrophosphates by cyclodextrins[a, b]

Substituent ($R_1 = R_2$ in (11))	Acceleration by cyclodextrin[c, d]		
	α-CD	β-CD	γ-CD
H	1.5	4.4	2.4
p-CH$_3$	1.1	9.2	6.0
p-Cl	>30	>200	>60

[a] From Hennrich, N., Cramer, F.: J. Amer. Chem. Soc. 87, 1121 (1965).
[b] pH 12.0, 40°.
[c] Ratio of the observed rate in the presence of 2.5×10^{-3} M cyclodextrin and Ca^{2+} to that in the presence of 2.5×10^{-3} M Ca^{2+} alone.
[d] CD stands for cyclodextrin.

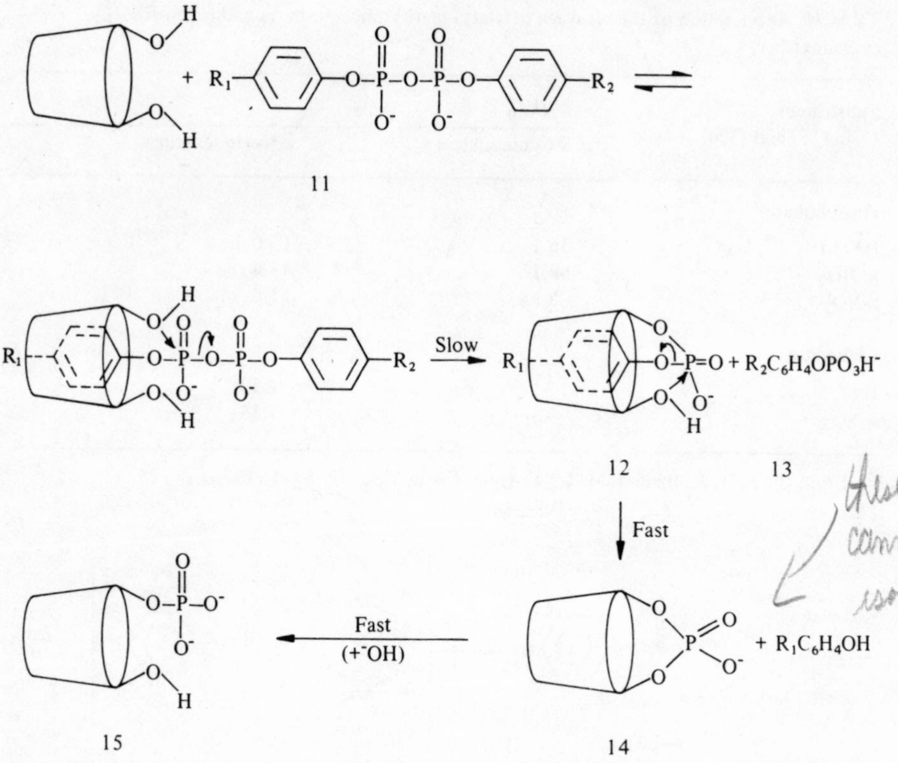

11

12   **13**   these cannot be isolated

| Fast

15   **14**

Scheme 3

even a slight shift in the position of the Ca^{2+} might have a considerable polarizing effect on the P—O—P bond.

Hydrolyses of diaryl methylphosphonates (16) and diaryl carbonates (17) are also accelerated by cyclodextrins (Table 20) [185]. Reactions proceed as shown in Scheme 4, which is similar to Scheme 3 proposed for the cyclodextrin-catalyzed hydrolyses of pyrophosphates. (The scheme for the hydrolyses of carbonates can be obtained simply

by replacing $-\overset{\overset{O}{\parallel}}{\underset{\underset{CH_3}{|}}{P}}-$ in Scheme 4 by $-\overset{\overset{O}{\parallel}}{C}-$).

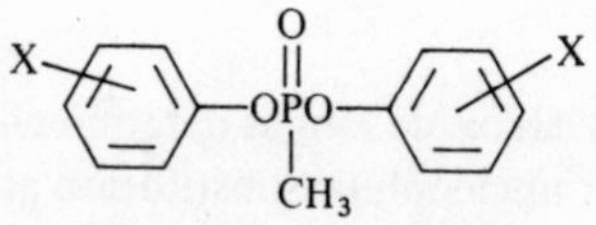

diaryl methyl phosphonate.
(16)

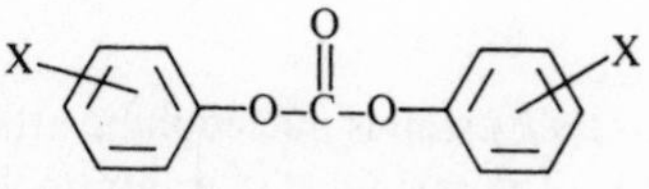

diaryl carbonate
(17)

Table 20. Acceleration of the cleavage of diaryl methylphosphonates and carbonates by cyclodextrins[a]

Substituent (X in (16) and (17))	k_2/k_{un}	
	α-Cyclodextrin	β-Cyclodextrin
Phosphonate		
H	35.1	16.0
m-NO$_2$	66.1	41.4
p-NO$_2$	8.35	11.5
Carbonate		
H	–	2.30
p-NO$_2$	–	7.45

[a] From Brass, H. J., Bender, M. L.: J. Amer. Chem. Soc. *95*, 5391 (1973).

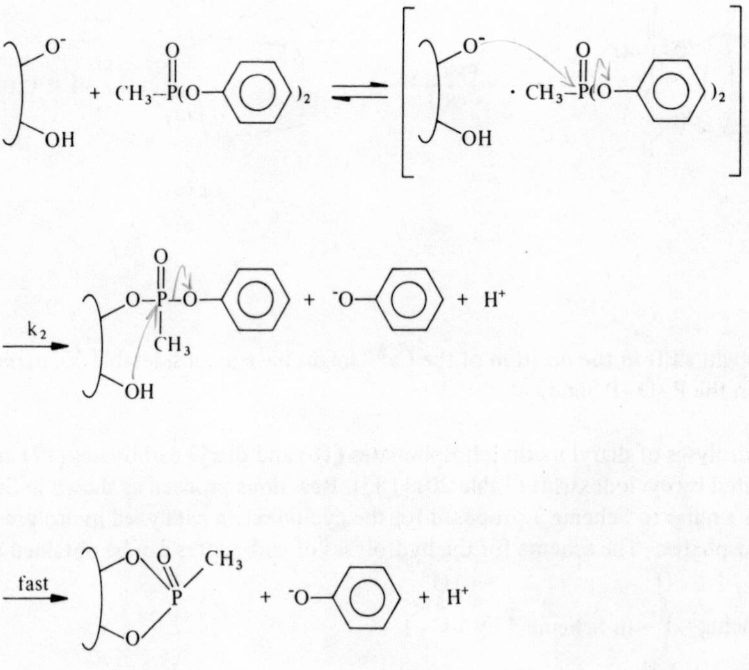

Scheme 4

The catalysis in the k_2 step is nucleophilic attack by alkoxide ion, as determined from the pH-k_2 profile. The second step is intramolecular nucleophilic substitution at phosphorus by the cyclodextrin anion. This intramolecular phosphorylation is so fast that it corresponds to catalysis by 10^3–10^4 M or more of intermolecular hydroxide ion.

In the second step, the intramolecular-nucleophilic substitution is accompanied by general base catalysis by a cyclodextrin hydroxyl ion.

Cleavage of dimethyl 4-nitrophenyl phosphate (18a) is also accelerated by cyclodextrin [248]. For example, at pH 10.0, 25 °C, the rate constant for the intracomplex nucleophilic attack by alkoxide ion of α-cyclodextrin on included (18a) is 2.4×10^{-2} min^{-1}, whereas the rate constant for the alkaline hydrolysis of free (18a) is 2.4×10^{-4} min^{-1}. However, the rate constant for the alkaline hydrolysis of included (18a) is 2.67×10^{-4} min^{-1}, which is almost equal to that for the alkaline hydrolysis of free (18a). Interestingly, the pK_a of the hydroxyl groups of α-cyclodextrin, kinetically determined from the dependence of the rate of the α-cyclodextrin-accelerated cleavage of (18a) on the concentration of hydroxide ion, is 11.4, which is considerably lower than the usual value for cyclodextrins (12–13) [51–53]. This was ascribed to the hydrogen bonding between the phosphonyl oxygen ($\geqslant$P=O) of (18a) included in α-cyclodextrin and one of the secondary hydroxyl groups of α-cyclodextrin.

In contrast to acceleration of cleavages of organophosphates by cyclodextrins, cleavages of monothiophosphate (18b) are retarded by β-cyclodextrin. The nucleophilic attack by the alkoxide ion of β-cyclodextrin on the included (18b) is almost perfectly suppressed. The rate of alkaline hydrolysis of the included (18b) is about 5 fold smaller than that of alkaline hydrolysis of free (18b). This result was interpreted in terms of the deep inclusion of (18b) in the cavity, resulting in the block of (18b) from the attack of alkoxide or hydroxide ion by the wall of the cyclodextrin. The electronegativity of sulfur (2.44) is less than that of oxygen (3.50), and the atomic radius of sulfur (1.27 Å) is larger than that of oxygen (0.6 Å). Therefore, the monothiophosphate group may be more hydrophobic than the corresponding phosphate group, and the former may be included more deeply in the β-cyclodextrin cavity than the latter. This presumption was supported by the fact that the inclusion complexes of β-cyclodextrin with monothiophosphates are more stable than those with the corresponding phosphates. (The K_d of (18a) with β-cyclodextrin is 6.3×10^{-3} M, whereas that of (18b) is 1.1×10^{-3} M.) On the other hand, phosphorus atoms of the phosphates (which are accelerated by cyclodextrins) are located in the vicinity of the secondary hydroxyl groups which are arranged around one edge of the torus of cyclodextrin.

(18a) X = O
(18b) X = S

There are other examples of the cyclodextrin-catalyzed hydrolyses of chiral organophosphates [53, 186]. However, these will be described in Chap. VII.

VI. Noncovalent Catalyses

Catalyses by cyclodextrins do not always involve formation of covalent intermediates. Instead, cyclodextrins can simply provide an apolar and sterically restricted cavity for the included substrate, which cavity can then serve as reaction medium, resulting in acceleration or deceleration of reaction. These kinds of catalyses are defined as non-covalent catalyses.

Noncovalent catalyses by cyclodextrins are attributable to:

a) microsolvent effects due to the apolar character of the cyclodextrin cavity

b) conformational effects due to the geometric requirements of inclusion

Hydrogen bonding between the substrate and the hydroxyl groups of the cyclodextrin can be considered to be a part of the microsolvent effect.

Noncovalent catalyses proceeds as illustrated in Scheme 5:

$$S + C \underset{k_{-1}}{\overset{k_1}{\rightleftharpoons}} S \cdot C \overset{k_2}{\longrightarrow} C + P_1 + P_2$$
$$\downarrow k_{un}$$
$$P_1 + P_2$$

Scheme 5

The designations are the same as in Scheme 1 for covalent catalyses.

Since Scheme 5 for noncovalent catalyses is formally identical with Scheme 1 for covalent catalyses except for the lack of a deacylation step, the rate constant, k_2, and the dissociation constant of the $S \cdot C$ complex, K_d, can be determined by use of either equation 2 or equation 3.

1. The Microsolvent Effect

A typical example of the microsolvent effect by cyclodextrins involves decarboxylation of anions of activated acids such as α-cyano and β-keto acids [158, 191]. These reactions proceed unimolecularly via rate-determining heterolytic cleavage of the carbon-carbon bond adjacent to the carboxylate group (Scheme 6) [249].

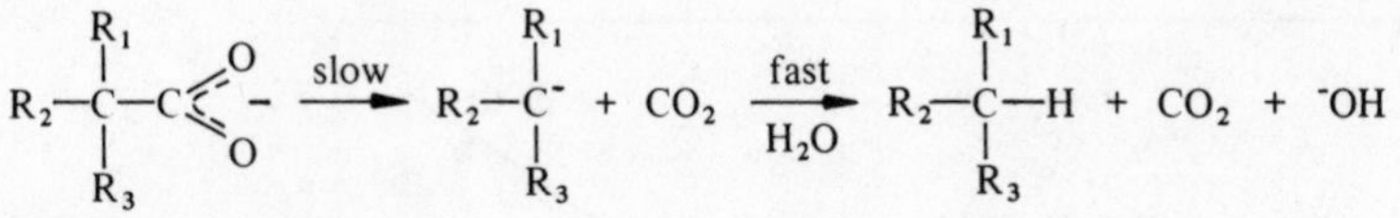

$$R_2-\underset{\underset{R_3}{|}}{\overset{\overset{R_1}{|}}{C}}-C\overset{O}{\underset{O}{\diagdown}}{}^{-} \xrightarrow{\text{slow}} R_2-\underset{\underset{R_3}{|}}{\overset{\overset{R_1}{|}}{C}}{}^{-} + CO_2 \xrightarrow[\text{H}_2\text{O}]{\text{fast}} R_2-\underset{\underset{R_3}{|}}{\overset{\overset{R_1}{|}}{C}}-H + CO_2 + {}^-OH$$

Scheme 6

The anionic decarboxylations are extremely solvent dependent, proceeding much faster in solvents of lower dielectric constant [250–252]. Cyclodextrin can accelerate the first step through the microsolvent effect, since the interior of the cavity has an apolar or ether-like atmosphere (Table 21) [158, 191]. Noncovalent catalysis rather than covalent catalysis were shown by

a) the catalytic rate constant, k_2, (as well as the uncatalyzed rate constant, k_{un}) follows the Hammett relationship (Fig. 13)

b) the activation parameters for cyclodextrin-catalyzed reactions are almost identical with a 2-propanol-H_2O mixture (Table 22) [191, 252]

c) the rate constant, k_2, is independent of pH and buffer catalyses [191]

d) the position of substituents (*ortho, meta,* and *para*) hardly affects the acceleration (k_2/k_{un}). In covalent catalyses, however, the stereo-specificity of k_2 is quite marked

The acceleration by cyclodextrin in decarboxylation is attributable to a decrease in activation enthalpy (smaller by 5.2 and 8.9 kcal/mole for *p*-chlorophenylcyano-acetate [191] and methyl *p*-chlorophenylcyanoacetate [190], respectively, than in water), which corresponds to a 10^3-10^6 fold acceleration. However, the decrease in activation enthalpy is partly compensated by a decrease in activation entropy, resulting in a small net acceleration by cyclodextrin.

Table 21. Kinetic parameters for the β-cyclodextrin-catalyzed decarboxylation of phenylcyano-acetic acid anions[a, b]

$R-C_6H_4CH(CN)CO_2^-$	k_{un} $(10^{-3} \text{sec}^{-1})$	k_2 $(10^{-3} \text{sec}^{-1})$	k_2/k_{un}	K_d (10^{-3}M)
R = *p*-CH$_3$O	0.0614	0.979	15.9	17.6
p-CH$_3$	0.119	1.51	12.7	15.7
m-CH$_3$	0.262	4.13	15.8	37.3
o-CH$_3$	0.374	4.48	12.0	67.8
H–	0.323	6.03	18.7	39.5
p-Cl	0.963	22.4	23.3	17.6
o-Cl	4.87	96.4	19.8	29.8
p-Br	1.21	20.1	16.6	8.54

[a] Reprinted with permission from Straub, T. S., Bender, M. L.: J. Amer. Chem. Soc. *94,* 8875 (1972). Copyright by the American Chemical Society.
[b] pH 8.6, 60,4°, I = 0.1 M.

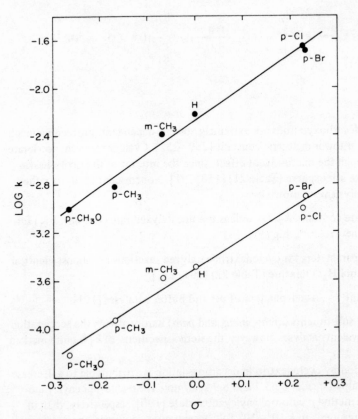

Fig. 13. Hammett $\rho\sigma$ correlations for the β-cyclodextrin catalyzed (●) and the uncatalyzed (○) decarboxylation of phenylcyanoacetic acid anions at pH 9.24. For the catalyzed reaction, the Hammett reaction constant $\rho = 2.72$. For the uncatalyzed reaction, $\rho = 2.44$. Reprinted with permission from Straub, T. S., Bender, M. L.: J. Amer. Chem. Soc. *94*, 8875 (1972). Copyright by the American Chemical Society

Table 22. Solvent dependence of the activation parameters in the decarboxylation of *p*-chloro-phenylcyanoacetic acid anion[a]

Solvent	$\Delta F^{\ddagger}$ [b] (kcal/mole)	E_a (kcal/mole)	$\Delta S^{\ddagger}$ (e. u.)
Water	25.0	31.3	19
β-Cyclodextrin cavity	22.5	26.1	10
57.7% (by weight) 2-Propanol-water	22.4	25.6	8.6
62.6% (by weight) 2-Propanol-water	22.3	25.3	8.2
74.0% (by weight) 2-Propanol-water	22.1	25.9	10.6

[a] Reprinted with permission from Straub, T. S., Bender, M. L.: J. Amer. Chem. Soc. *94*, 8875 (1972). Copyright by the American Chemical Society.
[b] 60.4°, I = 0.1 M.

Cramer and Kampe proposed another mechanism for the cyclodextrin-accelerated decarboxylation, which involves a general base-catalyzed nucleophilic attack by
a cyclodextrin hydroxyl oxygen atom at the carbonyl carbon of an acid anion. The
catalysis is allegedly assisted by hydrogen bonding between the keto or cyano group
and a cyclodextrin hydroxyl group in β-keto and α-cyano acid anion decarboxylation [189, 190]. However, in view of the more recent results, particularly the insen-
sitivity of the rate accelerations to the structure of the substrate, pH independence
of the rate, and absence of buffer catalysis [158, 191], a non-specific microsolvent
effect now seems more likely.

Another example of a microsolvent effect is the oxidation of α-hydroxyketones
to α-diketones. Here, cyclodextrins shift the keto-enol equilibria of the substrates
toward the more reactive enol form. Catalysis by cyclodextrins can be attributed to
this shift (Scheme 7) [192].

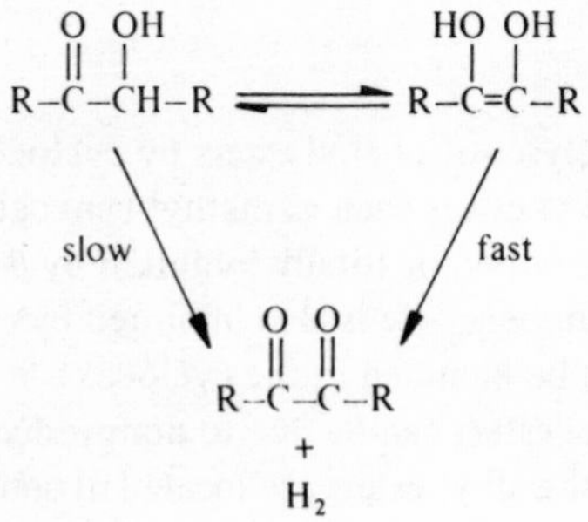

Scheme 7

The hydrolysis of 2,4-dinitrophenyl sulfate is accelerated by β-cyclodextrin by
18.7 fold [187]. This acceleration was attributed to the stabilization of the transition
state due to a microsolvent effect as well as the relief of strain induced by inclusion.
Stabilization of hydroperoxides by β-cyclodextrin was ascribed to hydrogen bonding
between the substrate and the cyclodextrin, and thus can be a kind of microsolvent
effect [200].

Trichlorphon (O,O-dimethyl 2,2,2-trichloro-1-hydroxyethylphosphonate) is a
contact and stomach poison insecticide. The first step of its decomposition consists
in the elimination of one molecule of HCl (Scheme 8). 1.6 Equivalent of β-cyclodex-
trin accelerates the first step of the decomposition by 1. 93 fold at pH 9.05, 30 °C.
Glucose does not have any effect on the rate of decomposition. The suggested mechanism involves inclusion complex formation of trichlorphon with β-cyclodextrin.
The interior of the cavity, which has a high electron density because of the lone pair
electrons of glucosidic oxygen atoms, can function as a "topochemical" base. Thus,
this is a microsolvent effect [196a].

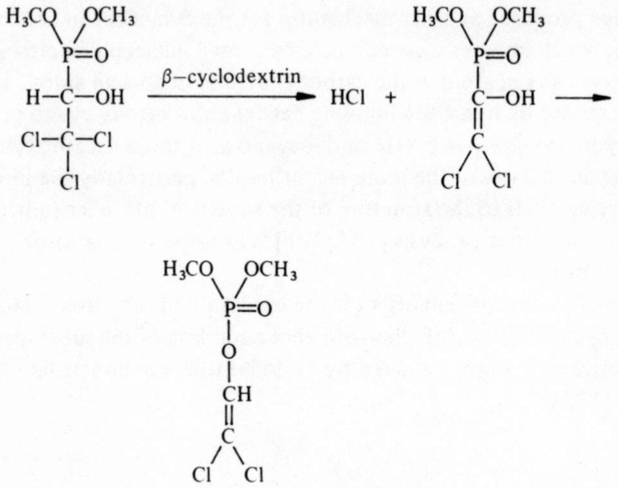

$$H_3CO\quad OCH_3$$
$$\diagdown \diagup$$
$$P{=}O$$
$$|$$
$$H{-}C{-}OH \xrightarrow{\beta-\text{cyclodextrin}} HCl\ + $$
$$|$$
$$Cl{-}C{-}Cl$$
$$|$$
$$Cl$$

$$H_3CO\quad OCH_3$$
$$\diagdown \diagup$$
$$P{=}O$$
$$|$$
$$C{-}OH \longrightarrow$$
$$\|$$
$$C$$
$$\diagup \diagdown$$
$$Cl\quad Cl$$

$$H_3CO\quad OCH_3$$
$$\diagdown \diagup$$
$$P{=}O$$
$$|$$
$$O$$
$$|$$
$$CH$$
$$\|$$
$$C$$
$$\diagup \diagdown$$
$$Cl\quad Cl$$

Scheme 8

In contrast to the accelerations of the hydrolyses of phenyl esters by cyclodextrins, described in Chap. V, the hydrolyses of alkyl esters such as methyl benzoates, ethyl benzoates, and ethyl *trans*-cinnamate are retarded or totally inhibited by β-cyclodextrin [51, 52, 152]. Hydrolysis of ethyl *p*-aminobenzoate is also inhibited by α-cyclodextrin (Table 23). All these substrates can be included in the cyclodextrin cavity [52]. The unreactivity of the complexed alkyl esters can be due to nonproductive binding in which the carbonyl carbon atoms of the alkyl esters are located at some distance from the secondary hydroxyl groups in the cyclodextrin cavity. Furthermore, the carbonyl carbon atoms in the cavity may be blocked from attack by hydroxide ion. This proposal is supported by the fact that the hydrolyses of atropine (19), ethyl *o*-aminobenzoate, and ethyl *m*-aminobenzoate, which can not be totally included in the α-cyclodextrin cavity because of steric hindrance, are slightly accelerated by α-cyclodextrin (Table 23) [52].

Table 23. The effect of α- and β-cyclodextrins on the hydrolyses of some alkyl esters[a, b]

Substrate	Rate effect of cyclodextrin[c]	
	α-Cyclodextrin	β-Cyclodextrin
Ethyl *p*-aminobenzoate	0.73	0.42
Ethyl *o*-aminobenzoate	1.58	0.90
Ethyl *m*-aminobenzoate	1.72	0.83
Atropine (19)[d]	1.60	0.63

[a] From Chin, T.-F., Chung, P.-H., Lach, J. L.: J. Pharm. Sci. 57, 44 (1968).
[b] pH 10.0, 70 °C.
[c] Ratio of the rate constant in the presence of 0.5% cyclodextrin to that in its absence.
[d] pH 10.0, 35 °C.

```
H₂C———CH———CH₂      CH₂OH
 |       |       |        |
 |      NCH₃   CH—OOC—CH
 |       |       |        |
H₂C———CH———CH₂      C₆H₅
```

(19)

Alternatively, the unreactivity of complexed alkyl esters is attributable to an unfavorable partitioning of the tetrahedral intermediate. Attack by an alkoxide ion of the cyclodextrin on the carbonyl carbon atom of the substrate leads to the formation of a tetrahedral intermediate having both cyclodextrin alkoxide ion and alkoxide ion derived from the alkyl ester as potential leaving groups. The tetrahedral intermediate will preferentially revert to reactants, since the cyclodextrin alkoxide ion is better a leaving group (lower pK_a) than the alkoxide ion derived from the alkyl ester.

Cyclodextrins can be matrices for stabilizing the radical formed by hydrogen atom addition to a benzene guest during γ-irradiation. Signals for cyclohexadienyl radicals trapped in cyclodextrin were observed by electron spin resonance spectroscopy, at 293 °K after cyclodextrin-benzene complexes were treated with γ-irradiation at 77 °K. Radicals are stabilized in the cyclodextrin cavity as well as in channels formed by cyclodextrin molecules. Radicals in the cavity are more stable than in the channel [253].

2. The Conformational Effect

Since the cavities of cyclodextrins are spatially restricted, cyclodextrins can include one conformational isomer of a substrate as a guest more favorably than other conformational isomers. When the conformer, which is more favorably included, is more reactive than the other conformers, cyclodextrins exhibit acceleration, and *vice versa*.

Conversion from (20) to (21) through intramolecular acyl migration is accelerated (6 fold) by α-cyclodextrin through this conformational effect (Scheme 9, Table 24) [188]. The reaction proceeds through intramolecular nucleophilic attack by the hydroxyl group on the carbonyl group [254], which is more probable in (20b) than in (20a). The acceleration by α-cyclodextrin may be attributed to an increase in activation entropy ($\Delta\Delta S^{\ddagger}$ = 4.3 e.u.); there is no appreciable change in activation enthalpy ($\Delta\Delta H^{\ddagger}$ = 0.2 kcal/mole). These values are consistent with the freezing of a rotational degree of freedom [255–258]. Thus, a portion of the free energy gained from the formation of an inclusion complex is used to increase the population of the reactive isomer (20b). This proposal was supported by stronger binding of (22), which has rigid structure and which is a transition state analog between (20) and (21), with α-cyclodextrin (K_d = 1.22 x 10^{-2} M) than (20) (K_d = 4.78 x 10^{-2}M). This is again an entropy effect. In contrast to the acceleration of the conversion of (20) to (21) by α-cyclodextrin, β-cyclodextrin retards this reaction by about 5 fold. β-Cyclodextrin is different from α-cyclodextrin only in the size of the cavity. This result is in agree-

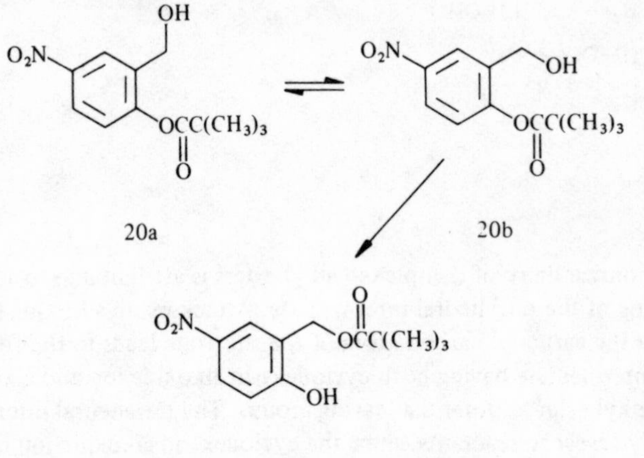

20a   20b

21

Scheme 9

Table 24. Activation parameters for the conversion of (20) to (21)[a, b]

Catalyst	k_2 (or k_{un}) $(10^{-2} \text{ sec}^{-1})$	K_d (M)	$\Delta H^{\ddagger}$ (kcal/mole)	$\Delta S^{\ddagger}$ (e. u.)
None	2.2	–	12.9	–22.6
α-Cyclodextrin	16	4.78×10^{-2}	13.1	–18.3
β-Cyclodextrin	0.41	9.60×10^{-4}	–	–

[a] From Griffiths, D. W., Bender, M. L.: J. Amer. Chem. Soc. *95*, 1679 (1973).
[b] pH 6.81, 25.2 °C.

ment with the proposal of a conformational effect, as are the entropy effects, since catalysis is definitely governed by the geometry of the fit of the substrate to the cavity.

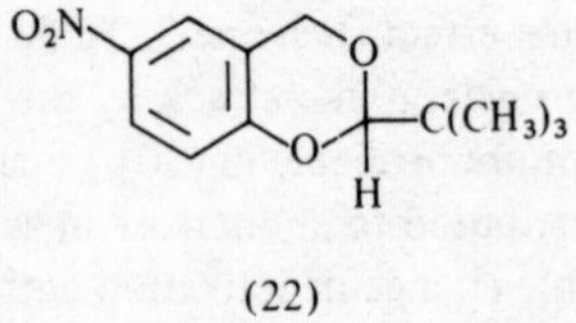

(22)

Another example of acceleration by cyclodextrin attributable to a conformational effect is the decarboxylation of β-keto acids (neutral form) [158]. As described in the earlier part of this Chapter, accelerations by α- and β-cyclodextrins of the decarboxylation of ions of β-keto acids are due to a microsolvent effect. In fact, transfer of these ions from aqueous solution to a 2-propanol-water or a dioxane-water solution can cause acceleration of their decarboxylation of almost same order of magnitude as shown by cyclodextrins. However, β-cyclodextrin exhibited a considerably larger acceleration in the decarboxylation of benzoylacetic acids (neutral form) than expected only from the microsolvent effect (Table 25). The ratio, k_2/k_{un}, for β-cyclodextrin is 5–6, whereas the rate in 50% dioxane-water solution is only about 2 times that in water. Besides, α-cyclodextrin showed an inhibitory effect in these reactions. Such a drastic difference between α-cyclodextrin and β-cyclodextrin can not be expected, if the effects of the cyclodextrins were a microsolvent effect. Consequently, accelerations by β-cyclodextrin in decarboxylations of β-keto acids (neutral form) were attributed to preferential inclusion of the more reactive conformer of the β-keto acid. The decarboxylation of un-ionized β-keto acids proceed through a cyclic transition state (23) (Scheme 10). β-Cyclodextrin should favorably include this cyclic ground state conformer. Inhibition by α-cyclodextrin might result from some interactions of the cyclic conformation with the wall of the cavity.

Table 25. Rate constants in the decarboxylation of unionized benzoylacetic acids[a, b]

Phenyl Substituent	Rate constant (10^{-3} sec^{-1})			
	β-cyclodextrin complex	Water	Dioxane-water (50% by volume)	Benzene
H	5.90	0.950	2.17	0.256
m-Cl	5.20	0.939	–	–
p-CH$_3$	6.04	0.967	–	0.388
p-NO$_2$	–	1.066	1.65	–

[a] From Straub, T. S., Bender, M. L.: J. Amer. Chem. Soc. 94, 8881 (1972).
[b] 50.3 °C.

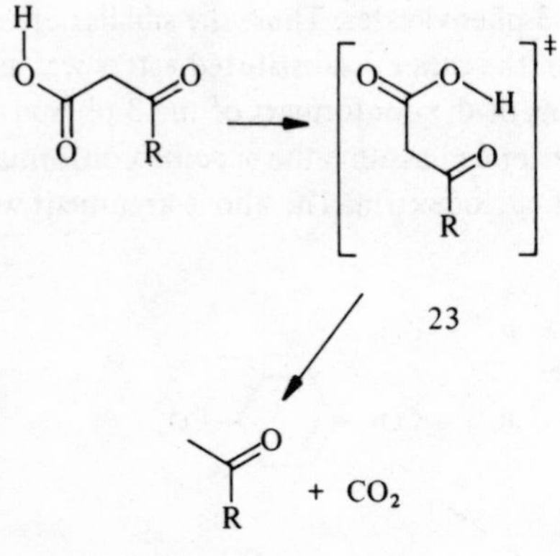

Scheme 10

Intramolecular carboxylate ion attack in mono-p-carboxyphenyl esters of 3-substituted glutaric acids (24) is markedly depressed by β-cyclodextrin through a conformational effect (Table 26) [201]. The 3-methyl, 3,3-dimethyl, and 3-isopropyl glutarate esters (24) form complexes with β-cyclodextrin which show little or no

Table 26. Deceleration of the hydrolyses of mono-*p*-carboxyphenyl esters of 3-substituted glutaric acids (24) by β-cyclodextrin[a, b]

3-Substituent	k_{un} $(10^{-4}\,\mathrm{sec}^{-1})$	k_2 $(10^{-4}\,\mathrm{sec}^{-1})$	k_2/k_{un}	K_d $(10^{-4}\,\mathrm{M})$
3,3-Dimethyl	20.1	0.40	0.02	4.6
3-Isopropyl	21.6	1.5	0.07	14.5
3-Methyl	4.54	0.04	0.009	32.8
3-Phenyl	1.96	0.47	0.24	13.1

[a] From van der Jagt, D. L., Killian, F. L., Bender, M. L.: J. Amer. Chem. Soc. *92*, 1016 (1970).
[b] pH 9.4, 30 °C, I = 0.2 M.

reactivity compared to the free esters. (24) can have two conformers; e.g. the more reactive conformer (24b) and less reactive conformer (24a) [255]. Of these two conformers, conformer (24b) can form an inclusion complex with cyclodextrin in such a way that the two carboxylate groups protrude from the top and the bottom of the cavity, and the rest of (24b) is inside of the cavity. However, conformer (24a) is too bulky to fit in the cavity. Consequently, the binding of (24) with cyclodextrin lowers the ground state energy of (24), resulting in deceleration (Scheme 11).

However, the 3-phenyl ester shows a smaller difference in reactivity between the free and complexed esters than the 3,3-dimethyl, 3-isopropyl, and 3-methyl esters do. A molecular model study indicates that the 3-phenyl ester can assume a conformation which allows an overlap of two benzene rings producing a structure which can fit partly into the β-cyclodextrin cavity. α-Cyclodextrin, which has too small a cavity to accommodate the 3-phenyl ester, even in the above conformation, exhibited no effect of the intramolecular reaction of 3-phenyl ester. Thus, the smaller effect of β-cyclodextrin on the 3-phenyl ester than the other 3-substituted esters was ascribed to a smaller change in the populations of the conformers of the 3-phenyl ester in complex formation, since the 3-phenyl ester can assume the specific conformation described above even in the absence of cyclodextrin. The above argument was

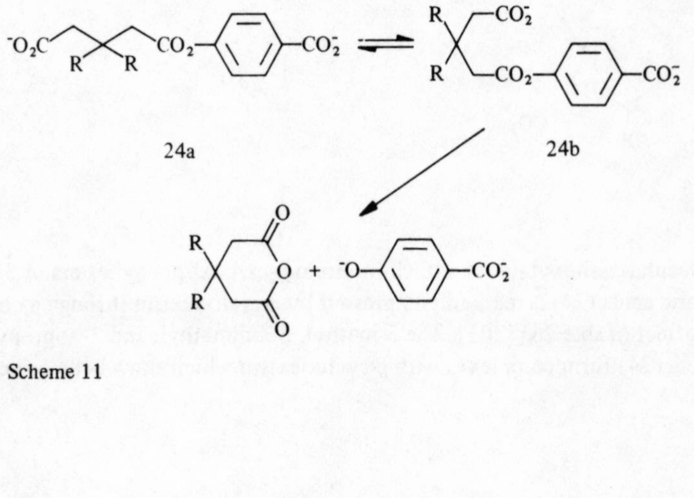

Scheme 11

supported by the fact that α-cyclodextrin depressed the intramolecular reactions of the 3,3-dimethyl and 3-methyl esters as β-cyclodextrin did [201].

Cyclodextrins retard the benzidine rearrangement of hydrazobenzene (Table 27) [202]. Complex formation probably restricts the large conformation change of the substrate necessary for reaching a "sandwich-type" intermediate (25) [259]. Alternatively, the apolar cavity may hinder the formation of the intermediate (25) through a microsolvent effect.

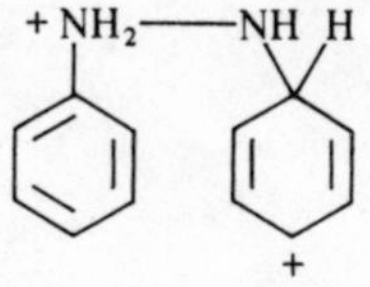

(25)

Although the mechanism has not been firmly established, the products of rearrangement of dibenzylmethylsulphonium (fluoroborate) (26) in alkaline solution are greatly affected by the presence of cyclodextrin [260]. In the absence of cyclodextrin (26), when treated with aqueous sodium hydroxide, generates the corresponding ylid (27), which then gives the Stevens rearrangement product (28) as the major product and the Sommelet rearrangement product (29) as the minor product ([29]/[28] = 0.36) (Scheme 12). However, equimolar amounts of β-cyclodextrin remarkably changes the product ratio ([29]/[28] = 1.5). Thus, the Sommelet rearrangement is more favorable than the Stevens rearrangement in the presence of cyclodextrin, which is the opposite to that found in its absence. The overall pseudo first-order rate constant for the rearrangement of (26) is slightly suppressed by cyclodextrin. Dependence of the rate constant on the cyclodextrin concentration showed that reaction takes place via complex formation of (26) with the cyclodextrin. Both the microsolvent effect and the conformational (or steric) effect of the cyclodextrin are possibilities. However, the latter is more plausible, since addition of methanol and ethylene glycol instead of β-cyclodextrin do not change the product ratio.

Table 27. Retardation of the benzidine rearrangement of hydrazobenzene by cyclodextrins[a, b]

Cyclodextrin	k_2/k_{un}	K_d $(10^{-3}$ M)
α-Cyclodextrin	0.099	30
β-Cyclodextrin	0.013	2.0

[a] From Matsui, Y., Mochida, K., Fukumoto, O., Date, Y.: Bull. Chem. Soc., Japan 48, 3645 (1975).
[b] pH 2.94, 25 °C, I = 0.5 M.

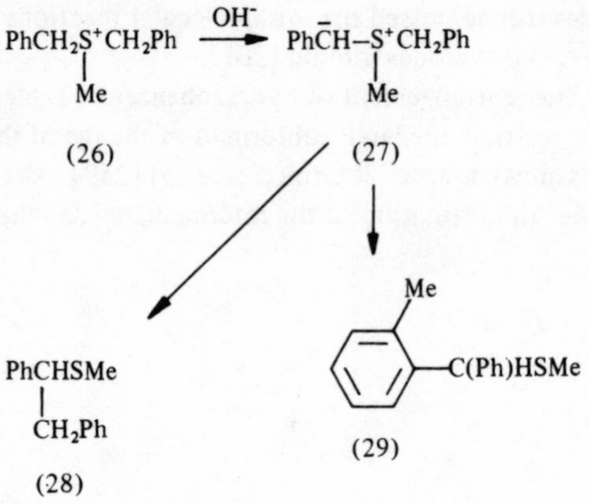

PhCH$_2$S$^+$CH$_2$Ph $\xrightarrow{\text{OH}^-}$ PhC̄H—S$^+$CH$_2$Ph

Me   Me

(26)   (27)

PhCHSMe

CH$_2$Ph

(28)

Me

—C(Ph)HSMe

(29)

Scheme 12

Cyclodextrin can also exhibit selectivity in the photochemical Fries rearrangement of phenyl esters (Scheme 13) [261]. Photoirradiation of phenyl acetate in the absence of cyclodextrin gave a mixture of *o*-hydroxyacetophenone and *p*-hydroxyacetophenone in a molar ratio of 1 : 1 together with the hydrolysis product, phenol. Addition of 0.45 equivalent of β-cyclodextrin considerably enhanced the production of the *p*-isomer, resulting in an *ortho-para* product ratio of 1 : 6.2. The conversion was increased about 3.5 fold by β-cyclodextrin. Interestingly, the formation of phenol was suppressed. The mechanism of the reaction in the presence of cyclodextrin, however, is not known yet.

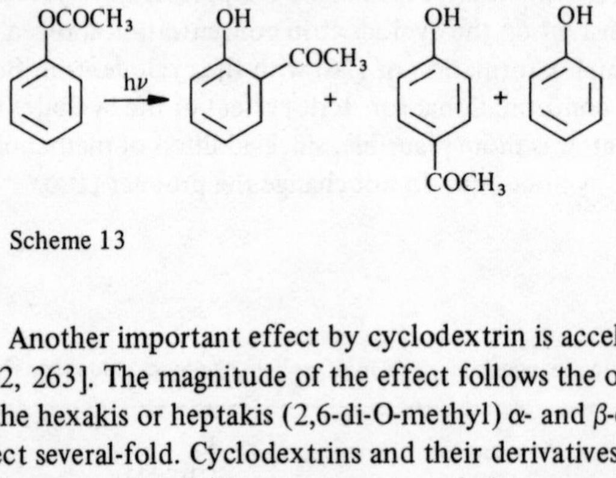

OCOCH$_3$ OH   OH   OH

$\xrightarrow{h\nu}$,COCH$_3$ +   +

                              COCH$_3$

Scheme 13

Another important effect by cyclodextrin is acceleration of fatty acid synthesis [262, 263]. The magnitude of the effect follows the order $\alpha > \beta > \gamma$. Methylation to the hexakis or heptakis (2,6-di-O-methyl) α- and β-cyclodextrins increased the effect several-fold. Cyclodextrins and their derivatives form complexes with palmitoyl coenzyme A. In these instances, cyclodextrins mimic the stimulatory effects of the mycobacterial polysaccharides on the activity of fatty acid synthetase [264, 265], although the mechanism of the catalysis is not established.

VII. Asymmetric Catalyses by Cyclodextrins

1. Selective Precipitation of D,L Compounds

The most simple usage of the chiral properties of cyclodextrins is selective precipitation of D,L compounds. D,L-isomers of most substrates have similar binding constants with cyclodextrins [108, 266]. However, D,L-isomers of certain substrates have appreciably different binding constants with cyclodextrins. Cyclodextrins can (partially) separate one enantiomer of these substrates from the other. That is, inclusion complexes, which are precipitated by addition of racemic substrates (in excess) to an aqueous cyclodextrin solution, contain more of one enantiomer than the other. Decomposition of the resulting inclusion complexes gives substrates which are rich in one enantiomer. By this method, chiral carboxylic acid esters [267], sulfoxides [268], O-alkyl alkylphosphinates [269], and O-alkyl alkylsulfinates [270] were resolved. The maximum optical purities, obtained after repeated precipitation with cyclodextrins followed by fractional crystallization, were 71.5, 84, and 70.2% for sulfoxides, phosphinates, and sulfinates, respectively.

2. Hydrolyses of Esters

The first observation of asymmetric specificity of cyclodextrins in catalysis was in the hydrolyses of ethyl mandelates; however, the asymmetric effect was quite small [163, 164]. A much larger asymmetric specificity was observed in the cyclodextrin-accelerated cleavage of 3-carboxy-2,2,5,5-tetramethylpyrrolidin-1-oxy m-nitrophenyl ester (30a), a substrate having an asymmetric carbon atom (like the mandelates) adjacent to the carbonyl group of the hydrolytically labile ester function [108, 110]. The hydrolysis of (30a) in the presence of cyclodextrin proceeds in the same way as that described for phenyl esters; i.e. binding of the substrate with cyclodextrin using the phenyl portion of the substrate, acylation of the cyclodextrin, and deacylation of the acyl-cyclodextrin. The cleavage of racemic (30a) in the presence of α-cyclodextrin was biphasic with a fast first step followed by a slower second step, because of the different rates at which the two complexed enantiomers are cleaved. The first and second phases, respectively, correspond to the cleavages of the (+) and (−) enantiomers. This assignment is based on the fact that both the catalytic rate constant, k_2, and the dissociation constant of the complex, K_d, determined from the dependence of the rates of the first phase on the cyclodextrin concentration, are equal to those determined by using the optically pure (+) enantiomer. As shown in Table 28, k_2 for the (+) enantiomer is 6.9 times larger than that for the (−) enantiomer. However, the K_d values are almost equal for the two enantiomers. Interestingly, in contrast to α-cyclodextrin, β-cyclodextrin exhibited no appreciable enantiomeric specificity. The loss of D,L-specificity upon increasing the size of cyclodextrin cavity indicates that D,L-specificity is derived from tight binding. The magnitude of enantiomeric specificity of k_2 shown by α-cyclodextrin is close to that (9.1 times) shown by α-chymotrypsin in the acylation step of the hydrolysis of the closely related ester, 3-carboxy-2,2,5,5-tetramethylpyrrolidinyl-1-oxy p-nitrophenyl ester (30b) [110].

30a: X=m−NO$_2$
30b: X=p−NO$_2$

Cyclodextrins, however, show no enantiomeric specificity in the deacylation step. The rate of hydrolysis of acyl-α-cyclodextrin derived from the (+) enantiomer of (30a) is equal to that derived from the (−) enantiomer. This can be associated with the fact that the nitroxide function is not included in the cyclodextrin cavity, which was shown by an electron spin resonance study. In the α-chymotrypsin-catalyzed hydrolysis of (30b), however, the acyl enzyme derived from the (+) enantiomer hydrolyzes 21 times faster than the acyl enzyme derived from the (−) enantiomer [110].

β-Cyclodextrin hydrolyzed L-acetylphenylalanine m-nitrophenyl ester 2 times faster than the D-enantiomer. Interestingly, β-cyclodextrin inhibited the hydrolysis of both L- and D-acetylphenylalanine p-nitrophenyl esters, indicating the predominant role of steric factors in these catalyses [271].

Table 28. Enantiomeric specificity by cyclodextrins in the hydrolyses of 3-carboxy-2,2,5,5-tetramethylpyrrolidin-1-oxy nitrophenyl esters (30a, 30b)[a,b]

Substrate	Catalyst	K_d $(10^{-2}$ M)	Rate constant (sec^{-1})	
			$k_{acylation}$ $(10^3\,k_2)$	$k_{deacylation}$ $(10^6\,k_3)$
+ (30a)	α-Cyclodextrin	1.3[c]	22[c]	8[c]
− (30a)	α-Cyclodextrin	1.3[c]	3.2[c]	8[c]
± (30a)	β-Cyclodextrin	0.075[d]	6.9[d]	32[e]
+ (30b)	α-Chymotrypsin	0.041[f]	370[f]	5,200[f]
− (30b)	α-Chymotrypsin	0.051[f]	41[f]	250[f]

[a] From Flohr, K., Paton, R. M., Kaiser, E. T.: J. Amer. Chem. Soc. 97, 1209 (1975).
[b] 25 °C. [c] pH 8.6. [d] pH 9.7. [e] pH 9.6. [f] pH 7.0.

3. Cleavage of Organophosphates

A much larger D,L-specificity was exhibited by cyclodextrin in the cleavage of chiral organophosphates such as isopropyl methylphosphonofluoridate (Sarin) (31) [186] and isopropyl p-nitrophenyl methylphosphonate (32) [53] (Table 29). The α-cyclo-dextrin-accelerated cleavage of these organophosphates proceed through nucleophilic attack of a secondary hydroxyl group of the cyclodextrin on the phosphorus atom, resulting in a phosphonylated α-cyclodextrin and hydrogen fluoride or p-nitrophenol. Thus, the catalyses are covalent catalyses.

$$
\begin{array}{cc}
\overset{\displaystyle O}{\underset{\displaystyle OCH(CH_3)_2}{CH_3-\overset{\|}{\underset{|}{P}}-F}} &
\overset{\displaystyle O}{\underset{\displaystyle OCH(CH_3)_2}{CH_3-\overset{\|}{\underset{|}{P}}-O-\langle\!\langle\ \rangle\!\rangle-NO_2}} \\
(31) & (32)
\end{array}
$$

In the α-cyclodextrin-accelerated cleavage of (31) the catalytic rate constant, k_2, for the (R)–(–) enantiomer is 35.6 times larger than that for the (S)–(+) enantiomer. This difference arises from the stereospecificity of the inclusion complexes, since the (S)–(+) enantiomer, which is less accelerated, binds to α-cyclodextrin more strongly than the (R)–(–) enantiomer, which is more accelerated [186, 266]. It was proposed that the stereospecific interaction(s) of the included enantiomers with the hydroxyl groups at the asymmetric C-2 atom and/or the asymmetric C-3 atom of the cyclodex-trin in the inclusion complexes govern this asymmetric catalysis.

On the other hand, the acceleration of the cleavage of the (S)–(+) enantiomer of (32) by α-cyclodextrin is too small to permit an accurate evaluation of the values of k_2 and K_d, although the cleavage of the (R)–(–) enantiomer of (32) is accelerated by α-cyclodextrin by 25 fold. Thus, the D,L-specificity is very large also in this reaction. The cyclodextrin-asymmetric phosphate system can be a simple model for the inhibi-tion of cholinesterases by organophosphorus compounds.

Table 29. Enantiomeric specificity by α-cyclodextrin in the cleavages of isopropyl methyl-phosphonofluoridate (31) and isopropyl p-nitrophenyl methylphosphonate (32)

Substrate	Enantiomer	k_{un} $(10^{-4}\ sec^{-1})$	k_2 $(10^{-4}\ sec^{-1})$	K_d $(10^{-2}\ M)$
31[a]	(R)–(–)	3.3	517	4.0
	(S)–(+)	3.3	14.5	0.60
32[b]	(R)–(–)	17	430	2.5
	(S)–(+)	17	–[c]	–[c]

[a] pH 9.0, 25 °C, from [186].
[b] pH 10.0, 25 °C, from [53].
[c] Acceleration by α-cyclodextrin is so small that the values of k_2 and K_d can not be accurately determined.

The larger enantiomeric specificity in the cleavages of organophosphates than in the cleavage of esters [107, 108] is attributable to the fact that the reaction takes place directly at the asymmetric center (phosphorus atom), whereas in the cleavage of esters it occurs at the carbonyl carbon atom, which is adjacent to the asymmetric center [272].

4. Addition and Oxidation Reactions

Addition of hydrogen cyanide to o- and p-chlorobenzaldehyde in the presence of α-cyclodextrin led to optically active α-hydroxynitriles ($[\alpha]_D^{25} = +0.2°$) and hence mandelic acids ($[\alpha]_D^{25} = -0.4°$) (Scheme 14). The reaction was carried out in water, the otherwise insoluble aldehydes being solvated by the α-cyclodextrin, which thus reacted in the asymmetric cavity. Addition of hydrogen cyanide to unsubstituted benzaldehyde caused no measurable optical activity in the products, indicating the importance of the stereochemistry of the inclusion complexes in asymmetric catalysis [163, 164].

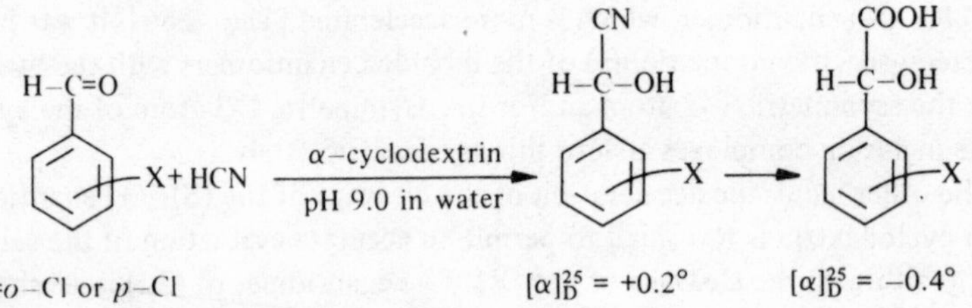

$X = o-Cl$ or $p-Cl$ $[\alpha]_D^{25} = +0.2°$ $[\alpha]_D^{25} = -0.4°$

Scheme 14

Furthermore, when 2,2'-dichlorobenzoin was 50% oxidized by molecular oxygen to 2,2'-dichlorobenzil in the presence of α-cyclodextrin, the remaining benzoin was optically active ($[\alpha]_D^{25} = + 0.37°$) [163, 164].

VIII. Improvement by Covalent and Noncovalent Modification

1. Acceleration of the Deacylation Step

As mentioned in earlier Chapters, cyclodextrins exhibit many interesting features such as rate effects, stereospecificity, enantiomeric specificity, etc., in organic reactions. However, there is one serious hortcoming of cyclodextrins as catalysts. That is the hydrolysis rates of acyl-cyclodextrins, intermediates in covalent catalysis of ester hydrolysis, are very slow, thus preventing cyclodextrins from acting as true catalysts. To accelerate the deacylation step, the addition of amines as nucleophiles were examined.

It was found that 1,4-diazabicyclo(2,2.2)octane (33), triethylamine, quinuclidine, piperidine, n-butylamine [273], 6-nitrobenzimidazole, and 5-nitrobenzimidazole [235] accelerate the hydrolyses of α-cyclodextrin- and β-cyclodextrin-*trans*-cinnamates and α-cyclodextrin-acetate (Table 30). Catalyses by these amines occur via inclusion complex formation in the cavity of the acyl-cyclodextrins. No relationship was observed between the catalytic activity of the amine and its complex formation constant. Furthermore, the catalytic activity is not a linear function of the pK_a of the amine. The geometry of the complex between the acyl-cyclodextrin and the amine should govern the catalytic process. The high catalytic activity of (33) can be also ascribed to the presence of two basic centers, which increases the frequency factor for reaction [273].

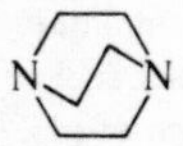

(33)

The logarithm of rate constant of hydrolysis of β-cyclodextrin-*trans*-cinnamate catalyzed by (33), (k_c), increases with pH with a slope of 1.0 up to pH 13.6. The logarithm of rate constant for its spontaneous hydrolysis in the absence of amines (k_{un}) also increased linearly with pH up to pH 11.2. Above pH 11.2, however, the slope gradually decreased with pH. As shown in Fig. 14, the ratio (k_c/k_{un}) remains rather constant at about 6 below pH 11.5. Above pH 11.5, however, the ratio drastically in creased with pH, attaining a value of 57 at pH 13.6 [273].

65

Table 30. The acceleration of the hydrolysis of acyl-cyclodextrins by several amines[a]

Acyl-Cyclodextrin	Catalyst	k_c/k_{un}	K_d[b] $(10^{-2}$ M)	pK_a of Amine
β-Cyclodextrin-*trans*-cinnamate	1.4-Diazabicyclo[2,2,2]octane (33)	6.7	12	8.19
	Triethylamine	2.4	38	10.78
	Quinuclidine	4.7	9.2	10.95
	Piperidine	3.3	9.4	11.28
	n-Butylamine	5.0	43	10.28
	Diisobutylamine	0.0	–	10.91
	6-Nitrobenzimidazole[c,d]	1.93	2.64	–
	5-Nitrobenzimidazole[c,d]	1.68	1.63	–
α-Cyclodextrin-*trans*-cinnamate	1,4-Diazabicyclo-[2,2,2]octane (33)	3.6	35	8.19
	6-Nitrobenzimidazole[c,d]	2.02	2.09	–
α-Cyclodextrin-acetate	6-Nitrobenzimidazole[c]	3.40[e]	–	–

[a] From Komiyama, M., Bender, M. L.: Proc. Nat. Acad. Sci. USA *73*, 2969 (1976); 20 °C, pH 12.0, unless otherwise noted.
[b] The dissociation constant of the complex between acyl-cyclodextrin and catalyst.
[c] From Kurono, Y., Stamoudis, V., Bender, M. L.: Bioorg. Chem. *5*, 393 (1976).
[d] 50 °C, pH 7.35.
[e] The ratio of the rate in the presence of 0.04 M catalyst to that in its absence at pD 8.2, room temperature.

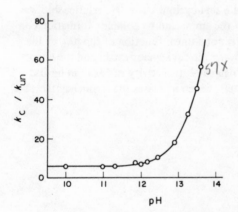

Fig. 14. The ratio (k_c/k_{un}) for the hydrolysis of β-cyclodextrin-*trans*-cinnamate catalyzed by 1.4-diazabicyclo[2.2.2]octane(33) at 20° as a function of pH. From Komiyama, M., Bender, M. L.: Proc. Nat. Acad. Sci. USA *73*, 2969 (1976)

In the presence of (33), cyclodextrin can be called a true catalyst, since the rate constant of the deacylation step, the rate-determining step, is larger than that of alkaline hydrolysis of the substrate. For example, in the hydrolysis of *m*-chlorophenyl benzoate, k_c, the rate constant for the hydrolysis of β-cyclodextrin-benzoate with (33) is 1.8×10^{-3} sec.$^{-1}$ at pH 10.6, 25°, which is 3.3 fold larger than the rate constant for alkaline hydrolysis of the substrate [273].

Cationic and hydrophobic polyelectrolytes such as poly-(4-vinyl-N-n-butylpyridinium bromide) and poly-(4-vinyl-N-ethylpyridinium bromide) accelerate the deacylation step as well as the acylation step in the cyclodextrin-accelerated hydrolysis of p-nitrophenyl indol-3-yl acetate (34) [274]. 6×10^{-3} M poly-(4-vinyl-N-n-butylpyridium bromide) accelerates the deacylation step (the hydrolysis of β-cyclodextrin-indol-3-ylacetate) about 4 fold. However, the diallyldiethylammonium chloride-sulphur dioxide copolymer (a cationic but less hydrophobic polyelectrolyte), sodium polystyrenesulphonate (an anionic polyelectrolyte), and sodium dodecyl sulphate (a surfactant) showed acceleration neither in the deacylation step nor in the acylation step. In a proposed mechanism for the acceleration of the deacylation by cationic and hydrophobic polyelectrolytes, hydroxide ions and the acyl-cyclodextrin are accumulated around the polyelectrolytes by hydrophobic and electrostatic interactions, which facilitate the deacylation. In a similar way, the substrate (34) and the cyclodextrin are accumulated around the polyelectrolyte, resulting in the acceleration of acylation. In the acylation step, it was shown that the hydrophobic interactions between the substrate and the polyelectrolyte are very important, since the cleavage of a less hydrophobic substrate, m-nitrophenyl acetate by cyclodextrin was retarded by the same polyelectrolytes, in contrast to the acceleration of the cleavage of (34) by cyclodextrin by them.

(34)

2. Better Models of Hydrolytic Enzymes

As described in Chap. V, covalent catalyses by cyclodextrins have many features similar to hydrolytic enzyme reactions. Especially:

a) an apolar cavity as the substrate binding site

b) an aliphatic hydroxyl group as the active site

c) the formation of a covalent intermediate

These mechanistic features are identical with those of serine proteases. Thus, it was expected that the introduction of an imidazolyl group, which is present at the catalytically active sites of serine proteases [275–278], would make cyclodextrins a better enzyme model.

The first attempt by Cramer and Mackensen [279, 280] at improvement of cyclodextrin catalysis of ester hydrolysis by the introduction of an imidazolyl group to cyclodextrin resulted in only a slight rate enhancement. They reacted 4-(or 5-)(chloromethyl)imidazole with β-cyclodextrin and obtained β-cyclodextrin derivatives with an average of two imidazolyl groups per molecule. This compound, however, exhibited only a 1.3 fold rate enhancement over the combination of two imidazole molecules and one native β-cyclodextrin molecule. The small effect of their modification can be attributed to the fact that the imidazolyl groups preferentially substituted the more reactive primary hydroxyl groups at the C-6 atoms of cyclodextrin rather than the less reactive secondary hydroxyl groups at the C-2 and C-3 atoms, which are effective in catalysis. In Cramer's compound, the imidazole groups introduced can not cooperate with the secondary hydroxyl groups operative in catalysis.

However, Iwakura and coworkers succeeded in selective modification of one of the secondary hydroxyl groups of cyclodextrin by a histamine group [281]. The key step of their synthesis was the selective tosylation of one secondary hydroxyl group, effected by the reaction of 10 equivalents of p-toluenesulfonyl chloride with α-cyclodextrin in pH 11 buffer solution at 25° for 1 hr. Here, one of the secondary hydroxyl groups at the C-3 positions of α-cyclodextrin can attack the sulfur atom of p-toluenesulfonyl chloride included in its cavity in the same way as reaction occurs in the cyclodextrin-accelerated hydrolyses of phenyl esters, giving the desired selective modification. Iodo-α-cyclodextrin, obtained by reaction of tosyl-α-cyclodextrin with NaI, was then converted to the final product, α-cyclodextrin-histamine (35). Modification at one of the secondary hydroxyl groups at the C-3 atoms was confirmed by ^{13}C-NMR spectroscopy. (35) accelerates the hydrolysis of p-nitrophenyl acetate 80 times more than α-cyclodextrin itself does and 6.3 times more than a mixture of α-cyclodextrin and histamine (Table 31). This modified cyclodextrin, which has a catalytic site containing both imidazolyl and hydroxyl groups together with a binding site (the apolar cavity of the cyclodextrin) as in the enzymes α-chymotrypsin, trypsin, elastase, and subtilisin. This modified cyclodextrin is a better enzyme model than cyclodextrin itself, since it shows rate acceleration around neutrality whereas rate acceleration by cyclodextrin shows up only in quite alkaline solution.

(35)

The effect of 2-benzimidazoleacetic acid (36) on the α-cyclodextrin-catalyzed hydrolysis of m-t-butylphenyl acetate was examined [240]. This noncovalently modified system was used as a probe of the "charge-relay" system exhibited by the exzymes α-chymotrypsin [282] and subtilisin [283], since this system contains imidazolyl and carboxyl groups of (36) in addition to the hydroxyl groups of α-cyclodextrin [284]. (36), which has both the imidazolyl and carboxyl groups in the same mole-

Table 31. Pseudo-first-order rate constants for the cleavage of p-nitrophenyl acetate in the presence of added catalysts[a, b]

Catalyst	Rate constant (10^{-4} sec^{-1})
None	0.789
α-Cyclodextrin	1.27
Histamine	12.1
α-Cyclodextrin + Histamine (1 : 1 mixture)	16.0
α-Cyclodextrin-Histamine (35)	101

[a] From Iwakura, Y., Uno, K., Toda, F., Onozuka, S., Hattori, K., Bender, M. L.: J. Amer. Chem. Soc. *97*, 4432 (1975).
[b] pH 8.37, 25 °C, I = 0.2 M.

cule, accelerated ester cleavage in the presence of α-cyclodextrin (Table 32). On the other hand, neither benzimidazole (which has only an imidazolyl group) nor 2-naphthaleneacetic acid (which has only a carboxyl group) exhibited measureable acceleration. (36) does not form a complex with α-cyclodextrin. The pH-rate constant profile showed that catalysis by the α-cyclodextrin-(36) system involves the combination of the carboxylate ion, the neutral imidazolyl group, and the alkoxide ion. Probably,

(36)

Table 32. Rate constants for the cleavage of the α-cyclodextrin-*m-t*-butylphenyl acetate complex by 2-benzimidazoleacetic acid (36) or benzimidazole[a, b]

Rate constant	2-Benzimidazoleacetic acid (36)	Benzimidazole
$10^6 k_C$ sec^{-1} [c]	4.2	4.2
$10^4 k_B$ M^{-1} sec^{-1} [d]	4.0	2.8
$10^4 k_{CB}$ M^{-1} sec^{-1} [e]	28	2.4

[a] From Komiyama, M., Breaux, E. J., Bender, M. L.: Bioorg. Chem. *6*, 127 (1977).
[b] pH 7.0, 25 °C, I = 0.2 M.
[c] k_C = rate constant of the α-Cyclodextrin-ester complex.
[d] k_B = rate constant of ester with base.
[e] k_{CB} = rate constant of the α-cyclodextrin-ester complex with base.

nucleophilic attack by the imidazolyl group on the included phenyl ester is assisted by the carboxylate and alkoxide ions. Thus the mechanism is apparently different from that shown by the "charge-relay system" in serine proteases. Serine proteases involve two proton transfers involving the carboxylate ion, the imidazolyl group, and the alkoxide ion. Serine proteases show nucleophilic attack by alkoxide ion whereas this process shows nucleophilic attack by imidazole.

The introduction of acetohydroxamic acid derivatives such as N-(N,N'-dimethyl-aminoethyl)acetohydroxamic acid (37a) [271], N-(4-imidazolemethyl)acetohydroxamic acid (37b) [271], and N-methylacetohydroxamic acid (37c) [285] on α-cyclodextrin exhibited a larger rate enhancement than either α-cyclodextrin itself or the acetohydroxamic acid derivatives. In these reactions, the cyclodextrin molecule provides the cavity as a binding site but does not function as a nucleophile.

$$OX$$

O OH
|| |
(37a) X = –CH$_2$C–N–CH$_2$CH$_2$N (CH$_3$)$_2$

O OH
|| |
(37b) X = –CH$_2$C–N–CH$_2$-[imidazole ring N⟋NH]

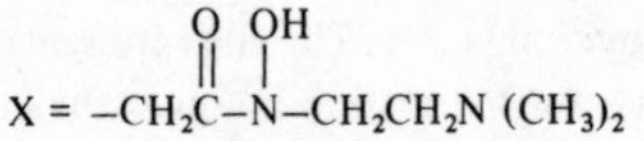

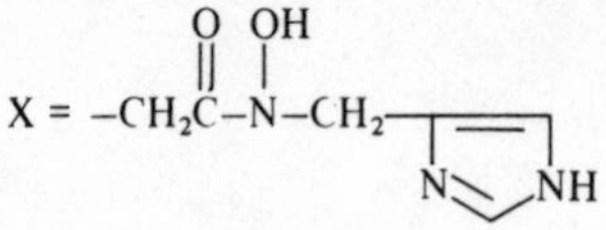

O
(37c) X = –[C=O]–N(–CH$_3$)–OH

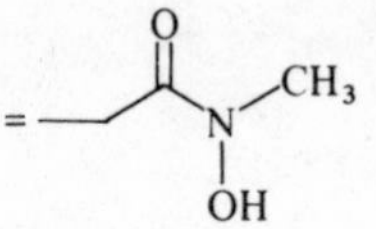

(37d) X = –CH$_2$–CH=N–OH

(37a) and (37b) were synthesized by the reactions of carboxymethyl-α-cyclodextrin methyl ester with 2-(N,N'-dimethylaminoethyl) hydroxylamine and 4-imidazole-methylhydroxylamine, respectively, in dimethyl sulfoxide. Carboxymethyl-α-cyclodextrin methyl ester was synthesized by reaction of α-cyclodextrin with sodium iodoacetate in the presence of sodium hydride in dimethyl sulfoxide, followed by esterification by diazomethane in N,N-dimethylformamide [271].

(37a) exhibited 2775, 2500, and 240 fold accelerations over native α-cyclodextrin in the cleavages of p-nitrophenyl thiolacetate, p-nitrophenyl acetate, and m-nitrophenyl acetate, respectively. (37a) exhibited a small D,L-selectivity in the cleavage of acetylphenylalanine m-nitrophenyl ester (L/D = 0.65), which is opposite D,L-selectivity to that found in the reaction by native α-cyclodextrin (L/D = 2.0). The reactions of (37a) with these substrates proceed through the acylation of the acetohydroxamic acid group attached to the cyclodextrin. The deacylation of acyl-(37a) is facilitated by intramolecular general base catalysis by the dimethylamino group [287]. Regeneration of (37a) was observed in the hydrolysis of acetic anhydride, where (37a) is a true catalyst.

(37b) showed 1525 and 1500 fold larger effects than native α-cyclodextrin in the cleavages of p-nitrophenyl thiolacetate and p-nitrophenyl acetate. Besides, (37b) accelerated the cleavage of acetylphenylalanine p-nitrophenyl ester, though native α-cyclodextrin inhibited the reaction [271].

(37c) was prepared by the reaction of carboxymethyl-α-cyclodextrin methyl ester with N-methylhydroxylamine in dimethyl sulfoxide. The periodate titration of carboxymethyl-α-cyclodextrin, a precursor of (37c), showed that substitution is at one of the C-2, C-3 secondary hydroxyl groups. Thus, the N-methylacetohydroxamic acid group is located at the secondary hydroxyl side of the cavity. As shown in Table 33, (37c) showed larger acceleration than both native α-cyclodextrin and N-methylmethoxyacetohydroxamic acid (38) (which does not have a cavity) in the cleavages of p-nitrophenyl acetate, m-nitrophenyl acetate, 2-hydroxy-5-nitro-α-toluene-sulfonic acid sultone, and p-t-butylphenyl chloroacetate. The rate of the cleavage of p-nitrophenyl acetate is proportional to the concentration of the ionized hydroxamate function. Thus, the reaction proceeds via nucleophilic substitution at the carbonyl carbon atom by the hydroxamate anion [285].

(37c) displayed a marked kinetic stereospecificity for p-nitro over m-nitrophenyl acetate (whereas α-cyclodextrin itself shows the reverse stereospecificity). In the complex, where the nitro group is included at the bottom of the cavity, the carbonyl group of the para-ester is located in proximity to the hydroxamate group. However, the carbonyl group of the meta-ester is located at a considerable distance from the

Table 33. Catalyses of ester hydrolyses by α-cyclodextrin modified by N-methylacetohydroxamic acid (37c)[a, b] and N-methyl methoxyacetohydroxamic acid (38)

Ester	Catalytic rate constant (M^{-1} sec^{-1})		
	37c	α-Cyclodextrin	N-methylmethoxyaceto-hydroxamic acid (38)[c]
p-Nitrophenyl acetate	72.4	0.08	4.01
m-Nitrophenyl acetate	7.3	0.9	1.30
2-Hydroxy-5-nitro-α-toluene-sulfonic acid sultone	45.2	0.25	0.65
p-t-Butylphenyl chloroacetate	7.0	0.06	2.6

[a] From Gruhn, W. B., Bender, M. L.: Bioorg. Chem. 3, 324 (1974).
[b] pH 7.95, 25 °C, I = 0.2 M.
[c]

$$CH_3O-\overset{\overset{\textstyle O}{\|}}{C}-\underset{\underset{\textstyle OH}{|}}{N}-CH_3$$

(38)

hydroxamate group in the complex. Thus, the *meta-para* specificity can be explained by proximity [285].

(37 d), which has an oxime group at one of the 2-O atoms of α-cyclodextrin, accelerates the cleavages of phenyl esters much more than native α-cyclodextrin [288]. The oximate anion attacks the substrate included in the cyclodextrin cavity, resulting in an acylated oxime with release of phenol. The hydrolysis of the acylated oxime is so slow that the modified cyclodextrin can not be repeatedly used. The rate of the intra-complex reaction between the oxime group and *p*-nitrophenyl acetate in the inclusion complex is equal to that of the intermolecular reaction of *p*-nitrophenyl acetate with 1.0 M D-glucose oxime, a monomeric analog of (37 d). In the cleavage of *m*-nitrophenyl acetate, however, the oxime group of (37 d) is equivalent to only 0.01 M external reagent. Molecular model studies show that the acyl portion of *p*-nitrophenyl acetate is located very close to the oxime group in the inclusion complex, whereas that of *m*-nitrophenyl acetate is located far from the oxime group. The proximity effect governs the reaction here in the same way as it governs the cleavages of the phenyl esters accelerated by native cyclodextrins. With the native cyclodextrins, *meta*-substituted esters are faster than their *para* counterparts whereas with the oxime-substituted cyclodextrins, just the opposite is the case, as seen above before in (37 a).

Siegel et al. [289] prepared β-cyclodextrin 2-, 3-, and 6-phosphoric acids (39 a, 39 b, 39 c). In (39 a) one of the primary hydroxyl groups at the C-6 atoms is modified; modification at the C-6 atom was obtained by the reaction of the cyclodextrin with 0.4 equivalent of diphenylphosphorochloridate, followed by catalytic hydrogenation. To obtain (39 a) and (39 b), where one of the secondary hydroxyl groups at the C-2 and C-3 atoms, respectively, is modified, β-cyclodextrin was treated with 0.27 equivalent of bis(*m*-nitrophenyl) phosphate in aqueous sodium hydroxide. Here, bis(*m*-nitrophenyl) phosphate, which is included in the cyclodextrin cavity, reacts with the secondary hydroxyl group of the cyclodextrin, which is followed by intramolecular cyclization with elimination of the phenol to form the 2,3-cyclic phosphate. Hydrolysis of the 2,3-cyclic phosphate gave a mixture of (39 a) and (39 b), which were separated by ion exchange chromatography.

All (39 a), (39 b), and (39 c) exhibited a larger catalytic effect in the exchange of *p-t*-butylphenacyl alcohol (40) tritiated in the methylene group than the native cyclodextrin at high pH (Table 34). Here, phosphate dianions in the modified cyclodex-

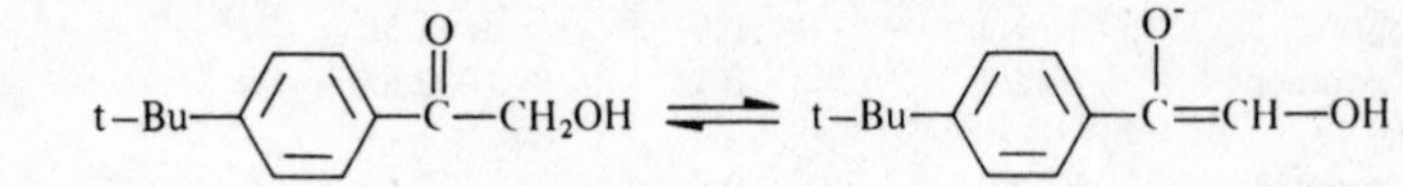

$$\text{t--Bu}\!-\!\!\left\langle\!\!\!\begin{array}{c}\\\end{array}\!\!\!\right\rangle\!\!-\!\!\overset{\overset{\displaystyle O}{\|}}{C}\!-\!CH_2OH \;\rightleftharpoons\; \text{t--Bu}\!-\!\!\left\langle\!\!\!\begin{array}{c}\\\end{array}\!\!\!\right\rangle\!\!-\!\!\overset{\overset{\displaystyle O^-}{|}}{C}\!=\!CH\!-\!OH$$

(40)   (41)

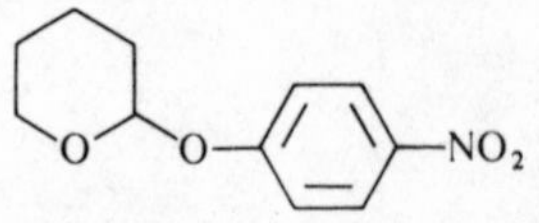

(42)

Table 34. The catalytic effects of (39a), (39b), and (39c)[a]

Reaction	Rate effect[b]		
	39a	39b	39c
Tritium exchange of (40)[c]	82.5	133	125
Hydrolysis of (42)[d]	1.5	5.2	1.0

[a] From Siegel, B., Pinter, A., Breslow, R.: J. Amer. Chem. Soc. *99*, 2309 (1977).
[b] The ratio of k_c by (39a), (39b), and (39c) to that by β-cyclodextrin.
[c] At pH 9.0.
[d] At pH 4.0, where the rate of the spontaneous hydrolysis of (42) is about 1,500 fold larger than the k_c for β-cyclodextrin.

trins catalyze the enolization of (40) to (41) as a general base. On the other hand, only (39b) of the three modified cyclodextrins showed net catalysis in the hydrolysis of p-nitrophenyl tetrahydropyranyl ether (42) at pH 4.0, where the phosphoric acid group of (39b) can function as a general acid.

In addition to modification by the introduction of catalytic groups, the introduction of certain (non-catalytic) groups on the cyclodextrin can make it a better enzyme model through improvement in the binding process rather than in the catalytic process. Compounds (43a) and (43b), which have several N-formyl groups at the bottom of the β-cyclodextrin torus (the primary alcohol group side), exhibit 10—20 times larger catalytic rate constants (k_2) in hydrolyses of phenyl esters than native β-cyclodextrin does [174]. However, the modified cyclodextrins have equal or smaller binding constants with phenyl esters than native cyclodextrin does (Table 35). Molecular models show that the alkyl groups in (43a) and (43b) cluster close to the bottom of the cavity, forming an apolar floor on the cavity. Thus, these flexibly capped cyclodextrins have a more apolar and more restricted cavity than the native cyclodextrin. The larger catalytic rate constants of (43a) and (43b) are probably due to more rigid and better binding of the substrates in the cavity. The weaker binding of modified cyclodextrins can be attributed to too shallow a cavity for the substrates to be sufficiently included. It is noteworthy that another substrate (1-adamantanecarboxylic acid) forms a complex with (43a) 20 times more strongly than with native β-cyclodextrin.

X X X X

(43a) X = —N—CH=O
       |
       CH_3

(43b) X = —N—CH=O
       |
       CH_2CH_3

Table 35. Accelerations of hydrolyses of phenyl esters by capped β-cyclodextrins (43a, 43b)[a, b]

Substrate	Accelerator	k_2/k_{un}	K_d $(10^{-4}$ M$)$
m-Nitrophenyl acetate	43a	660	51
	43b	1,140	260
	β-Cyclodextrin	64	53
m-t-Butylphenyl acetate	43a	3,300	4.6
	β-Cyclodextrin	365	1.95

[a] From Emert, J., Breslow, R.: J. Amer. Chem. Soc. 97, 670 (1975).
[b] pH 9.0, 25 °C, I = 0.2 M.

Cyclodextrin rigidly capped with a benzophenone chromophore

can be an effective donor of triplet energy to the included guests. The benzophenone chromophore of this capped cyclodextrin was selectively excited at 353 nm. The effective triplet-triplet energy transfer from the capped cyclodextrin (2.3×10^{-4} M) to 1-bromonaphthalene (5.0×10^{-4} M) was successfully observed at liquid nitrogen temperatures even in this considerably unfavorable (low) concentration for the transfer. On the other hand, an open-chain analog, benzophenone 4,4'-bis(carboxymethyl ester), did not show any appreciable energy transfer to 1-bromonaphthalene (2.4×10^{-3} M). However, no energy transfer was observed from the capped cyclodextrin to trisodium naphthalene-1,3,6-trisulfonate (5.0×10^{-3} M), which is a very hydrophilic and a poorly binding guest. Thus, the binding of the energy acceptor in the cavity of the capped cyclodextrin has an important role in the energy transfer process [148b].

In addition to the introduction of a noncatalytic apolar groups on cyclodextrin, the introduction of polar groups was also examined [290]. This can be important for the practical use of cyclodextrins, since the formation of inclusion complexes of native cyclodextrin with apolar substances often produces a water-insoluble complex. Amination of cyclodextrin selectively at the C-2 positions is also important for practical use, since aminated α-cyclodextrin is more stable against acid hydrolysis than native α-cyclodextrin [291].

There have been many reports on the modification of cyclodextrins in addition to those mentioned above [5, 40, 45, 238, 239, 292—298]. However, attention should be given to the numbers and positions of modified groups, since they should have considerable influence on catalyses by cyclodextrins. General organic procedures usually cause nonspecific modification of cyclodextrins, although the primary hydroxyl groups at the C-6 atoms are more reactive than the secondary hydroxyl groups at the C-2 and C-3 atoms. Thus, specific modification of cyclodextrins requires refined techniques, followed by separation of the correctly modified product from the incorrectly modified products by column chromatography.

Modification of only one of the secondary hydroxyl groups at the C-3 positions is carried out by reaction of 10 equivalents of *p*-toluenesulfonyl chloride with cyclodextrin in pH 11.0 buffer [281], as described before. On the other hand, reaction of 6–9 equivalents of *p*-toluenesulfonyl chloride with cyclodextrin in pyridine results in tosylation of all the primary hydroxyl groups at the C-6 positions without any tosylation of secondary hydroxyl groups [293]. When the reaction is carried out in pyridine for a short time (40 min), the mono-tosylated compound (modified at the primary hydroxyl group) is obtained together with unreacted and more highly substituted cyclodextrins, which can be removed by column chromatography [294]. One possible explanation for the different specificity in tosylation between that in pH 11.0 buffer solution (modification at the secondary hydroxyl group at the C-3 position) and that in pyridine (modification at the primary hydroxyl group(s) at the C-6 position(s)) is that the secondary hydroxyl groups are largely ionized and catalytically active in pH 11.0 buffer solution, but not in pyridine. Beside, in pyridine, *p*-toluenesulfonyl chloride, which can not effectively form complexes with cyclodextrins because of competitive binding by pyridine, reacts preferentially with the more reactive primary hydroxyl groups. Conversion of these specifically tosylated cyclodextrins to other specifically modified cyclodextrins can be effected by usual organic techniques.

Disubstituted cyclodextrins can be easily synthesized by use of a capped cyclodextrin (7) as a starting material [299]. (7) is activated at the bridgehead positions of the capping group toward the nucleophilic displacement, which results in easy disubstitution. For example, reaction of (7) with thiophenol at 50 °C for 24 hr. gave (44) in 20% yield. Two $-N_3$, $-N(C_2H_5)_2$, and $-SCH_2CH_2NH_2$ groups are introduced to cyclodextrin by similar procedure.

PhS   SPh

(44)

Recently, polymers containing cyclodextrins on their chains were synthesized [300, 301]. An acryloyl group was selectively introduced on one of the secondary hydroxyl groups by the reaction of β-cyclodextrin with *m*-nitrophenyl acrylate or its derivatives [281]. The acrylic monomers containing β-cyclodextrin were homopolymerized or copolymerized with other acrylic monomers by a radical initiator. The resulting polymers or copolymers exhibited slightly larger catalytic rate constants (about 2–3 fold) in the hydrolyses of *p*-nitrophenyl esters than β-cyclodextrin did itself [300, 302]. The high local concentration of catalytic hydroxyl groups of pendant β-cyclodextrins on the polymer chain may lead to multifunctional catalysis.

These cyclodextrin-containing polymers exhibited a larger enhancement of fluorescence of potassium 2-p-toluidinylnaphthalene-6-sulfonate than the parent cyclodextrin (see also p. 15). For example, poly(acryloyl-β-cyclodextrin) exhibited 23 fold greater enhancement than β-cyclodextrin. The polymer complex showed exclusively 2:1 stoichiometry, whereas the complex of β-cyclodextrin showed 1:1 stoichiometry at low concentrations of the cyclodextrin and 2:1 stoichiometry at high concentrations. Two cyclodextrin molecules on the polymer chain cooperate in binding one fluorescent molecule. It was proposed that the fluorescent enhancement caused by binding is ascribed to the restriction of intramolecular rotation in the rigid environment and/or to the absence of solvent relaxation [303].

3. Cyclodextrins Containing Metal Ions

Although metal ions have large catalytic effects [304] such effects are limited to substrates which can bind the metal ions. Thus, it was expected that the combination of catalytic ability of a metal ion and the binding ability of a cyclodextrin would give an excellent catalyst.

Using this hypothesis, modified α-cyclodextrins (45a) and (45b) were synthesized. Table 36 lists their catalytic ability in the hydrolysis of p-nitrophenyl acetate [305]. The observed rate constant, k_{obs}, refers to the first step of the catalysis, acetylation of the pyridine-carboxaldoxime (PCA) moiety in (45b), which is followed by metal ion-catalyzed hydrolysis of the PCA acetate [306]. (45b) is four times more reactive than is an equivalent concentration of the simple system without the cyclodextrin (PCA–Ni^{2+}), corresponding to a rate acceleration of greater than 10^3 over

Table 36. Pseudo first-order rate constants in the cleavage of p-nitrophenyl acetate catalyzed by (45a) and (45b)[a]

Catalyst (mM)	k_{obs} (min^{-1})
None	7.1×10^{-5} (25°)
45b (10.0)	19.3×10^{-2} (30°)
PCA–Ni^{2+} (10.0)	5.06×10^{-2} (30°)
45a–Ni^{2+} (5.0)	2.2×10^{-4} (55°)
α-CD + Ni^{2+}	2.2×10^{-4} (55°)

[a] From Breslow, R., Overman, L. E.: J. Amer. Chem. Soc. 92, 1075 (1970).
[b] pH 5.17, I = 1.0 M.

the uncatalyzed rate. The larger catalytic effect of (45b) than the PCA–Ni^{2+} system is attributable to the binding of the substrate by the cyclodextrin moiety of (45b). This was supported by the fact that 8-acetoxy-5-quinolinesulfonate, which does not fit in the cyclodextrin cavity, is only 57% as reactive toward (45b) as toward PCA–Ni^{2+}. The smaller difference than expected is associated with the freezing of several degrees of rotational freedom on approach of the PCA oxygen atom (the catalytic site) to the carbonyl carbon atom of the bound substrate in the inclusion complex.

The 2:1 complex of mono-(6-β-aminoethylamino-6-deoxy)-β-cyclodextrin (46) with Cu^{2+} accelerates the oxidation of furoin (47) (Table 37) [307]. The function of the amino group is presumably to complex the Cu^{2+}. However, β-cyclodextrin, 1,2-diaminoethane, the 2:1 diaminoethane-Cu^{2+} complex, and (46) exhibited slight accelerations of reaction. The rate constant catalyzed by the 2:1 (46)–Cu^{2+} complex is about 20 times that for the uncatalyzed reaction. Probably, the two cyclodextrins of the 2:1(46)–Cu^{2+} complex fix one furoin molecule with each of the two furan moieties complexed in one cavity. In this structure, the enolate anion of furoin, a reaction intermediate, is electrostatically stabilized by the Cu^{2+} ion. However, there is also the possibility of coordination of the enolate anion directly to the Cu^{2+} ion.

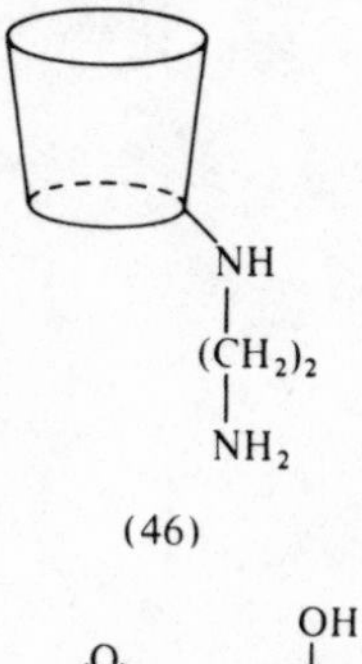

NH
|
(CH$_2$)$_2$
|
NH$_2$

(46)

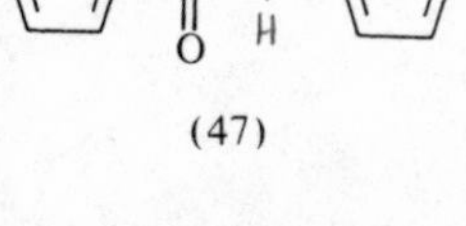

(47)

Table 37. Pseudo first-order rate constants in the oxidation of furoin (47)[a, b]

Catalyst (mM)	k_{obs} (min^{-1})
None	0.0572
β-Cyclodextrin (2.0)	0.0583
1.4-Diaminoethane (2.0)	0.0793
1.4-Diaminoethane (2.0) + Cu^{2+} (1.0)	0.0722
(46) (2.0) + Cu^{2+} (1.0)	0.337

[a] From Matsui, Y., Yokoi, T., Mochida, K.: Chem. Lett. *1976*, 1037.
[b] pH 10,5, 25°, [47] = 0.025 mM, [O$_2$] = 0.30 mM.

β-Cyclodextrins functionalized with polyamines (48a, 48b) strongly interact with metal ions such as Cu^{2+}, Zn^{2+}, and Mg^{2+}, and the resultant cyclodextrins flexibly capped by metal ions (49a, 49b, 49c) bind several anions with hydrophobic moieties much more strongly than the native cyclodextrin or (48a, b). Thus, (49b) binds adamantan-2-one-1-carboxylate better than β-cyclodextrin by 330 fold. Such binding enhancements were observed only for (49) in complexing with hydrophobic anions of the types $-CO_2^-$, $-SO_3^-$, and $-O^-$, whereas (48a, b) were only two to three times more effective than unsubstituted α-cyclodextrin. Binding enhancement was attributed to the cooperation of the interaction between the hydrophobic group and the cyclodextrin together with the interaction between the anion and the metal cation [299a].

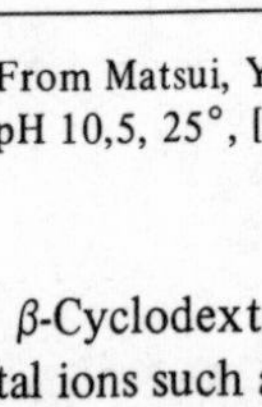

L

M^{2+}L

48a) L = $-NHCH_2CH_2NHCH_2CH_2NH_2$
48b) L = $-NHCH_2CH_2NHCH_2CH_2NHCH_2CH_2NH_2$

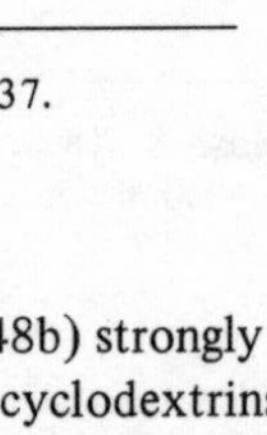

49a) M^{2+} = Cu^{2+}
49b) M^{2+} = Zn^{2+}
49c) M^{2+} = Mg^{2+}

IX. Conclusion

For decades, chemistry has been modeled on billiard ball reactions. However, the chemistry of the future will undoubtedly involve prior, probably stereospecific, complexing of reactants before reaction per se occurs. The cyclodextrins are the first organic compounds found that show the ability to complex other organic compounds. Their stereospecificity is limited, although they do exhibit stereospecific reactions. As noted above, they do catalyze many organic reactions, in some ways mimicing the action of enzymes. Just as enzymes do not engage in billiard ball chemistry, the cyclodextrins do not engage in billiard ball chemistry (usually). The fact that enzymes can achieve such high rates of reaction indicates that prior complexing followed by reaction must be the chemistry of the future.

More and more chemists and laboratories in many parts of the world are coming to this realization. This area of research is increasing in depth and breadth. It is hoped that this book will add something to this growing field.

References

1. Cramer, F.: Einschlußverbindungen. Berlin: Springer-Verlag 1954.
2. Griffiths, D. W., Bender, M. L.: Adv. Cat. *23*, 209 (1973).
3. Thoma, J. A., Stewart, L.: in Starch: Chemistry and Technology. Whistler, R. L., Paschall, E. F., (eds.), pp. 209–249. New York: Academic Press 1965.
4. Senti, F. R., Erlander, S. R.: in: Non-Stoichiometric Compounds. Mandelcorn, L., (ed.), pp. 588–601. New York: Academic Press 1964.
5. French, D.: Adv. Carbohyd. Chem. *12*, 189 (1957).
6. Cramer, F., Hettler, H.: Naturwissenschaften *54*, 625 (1967).
7. Bender, M. L., Komiyama, M.: in: Bioorganic Chemistry, Vol. 1, Chap. 2. VanTamelen, E. E. (ed.). New York: Academic Press 1977.
8. Bergeron, R. J., J. Chem. Educ. *54*, 204 (1977).
9. Saenger, W.: in: Environment Effects on Molecular Structure and Properties. Pullman, B. (ed.), pp. 265–305. Dordrecht/Holland, D. Reidel Publ. Co.
10. Hershfield, R., Bender, M. L.: J. Amer. Chem. Soc. *95*, 1376 (1972).
11. Cram, D. J., Cram, J. M.: Science *183*, 803 (1974).
12. Villiers, A.: Compt. Rend. Acad. Sci. Paris *112*, 536 (1891).
13. Schardinger, F.: Z. Unters. Nahrungs-, Genußmittel, Gebrauchsgegenstände 6, 865 (1903).
14. Schardinger, F.: Wien. Klin. Wochenschr. *17*, 207 (1904).
15. Schardinger, F.: Zentr. Bakteriol. Parasitenk. II, *29*, 188 (1911).
16. Cramer, F., Steinle, D.: Ann. Chem. *595*, 81 (1955).
17. Cramer, F., Henglein, F. M.: Chem. Ber. *91*, 308 (1958).
18. Freudenberg, K., Jacobi, R.: Ann. Chem. *518*, 102 (1935).
19. French, D., Levine, M. L., Pazur, J. H., Norberg, E.: J. Amer. Chem. Soc. *71*, 353 (1949).
20. Pulley, A. O., French, D.: Biochem. Biophys. Res. Commun. *5*, 11 (1961).
21. French, D., Pulley, A. O., Effenberger, J. A., Rougvie, M. A., Abdullah, M.: Arch. Biochem. Biophys. *111*, 153 (1965).
22. Carter, J. H., Lee, E. Y. C.: Anal. Biochem. *39*, 521 (1971).
23. Sundararajan, P. R., Rao, V. S. R.: Carbohyd. Res. *13*, 351 (1970).
24. Sato, M., Nakamura, N.: Japan Kokai Patent 74 92, 288; Chem. Abstr. *84*, 119962q (1976).
25. Beadle, J. B.: J. Chromatogr. *42*, 201 (1969).
26. Casu, B., Gallo, G. G., Reggiani, M., Vigevani, A.: Stärke *20*, 387 (1968).
27. Kondo, H., Nakatani, H., Hiromi, K.: Carbohyd. Res. *52*, 1 (1976).
28. Wiedenhof, N.: J. Chromatog. *15*, 100 (1964).
29. Hybl, A., Rundle, R. E., Williams, D. E.: J. Amer. Chem. Soc. *87*, 2779 (1965).
30. Hamilton, J. A., Steinrauf, L. K., VanEtten, R. L.: Acta Cryst. *B24*, 1560 (1968).
31. Takeo, K., Kuge, T.: Agr. Biol. Chem. *34*, 568 (1970).
32. Takeo, K., Kuge, T.: Agr. Biol. Chem. *34*, 1787 (1970).
33. Freudenberg, K., Blomqvist, G., Ewald, L., Soff, K.: Chem. Ber. *69B*, 1258 (1936).
34. French, D., Rundle, R. E.: J. Amer. Chem. Soc. *64*, 1651 (1942).
35. French, D., Knapp, D. W., Pazur, J. H.: J. Amer. Chem. Soc. *72*, 5150 (1950).
36. French, D., McIntire, R. L.: J. Amer. Chem. Soc. *72*, 5148 (1950).

37. James, W. J., French, D., Rundle, R. E.: Acta. Cryst. *12*, 385 (1959).
38. Rao, V. S. R., Foster, J. F.: J. Phys. Chem. *67*, 951 (1963).
39. Glass, C. A.: Can. J. Chem. *43*, 2652 (1965).
40. Casu, B., Reggiani, M., Gallo, G. G., Vigevani, A.: Tetrahedron *24*, 803 (1968).
41. Casu, B., Reggiani, M., Gallo, G. G., Vigevani, A.: Carbohyd. Res. *12*, 157 (1970).
42. Takeo, K., Kuge, T.: Agr. Biol. Chem. *34*, 1416 (1970).
43. Casu, B., Reggiani, M.: J. Polym. Sci. *C, 7*, 171 (1964).
44. Rao, V. S. R., Foster, J. F.: Biopolymers. *1*, 527 (1963).
45. Cramer, F., Mackensen, G., Sensse, K.: Chem. Ber. *102*, 494 (1969).
46. Colson, P., Jennings, H. J., Smith, I. C. P.: J. Amer. Chem. Soc. *96*, 8081 (1974).
47. Takeo, K., Hirose, K., Kuge, T.: Chem. Lett. *1973*, 1233.
48. Usui, T., Yamaoka, N., Matsuda, K., Tuzimura, K., Sugiyama, H., Seto, S.: J. Chem. Soc., Perkin Trans. *1*, 2425 (1973).
49. Casu, B., Reggiani, M., Gallo, G. G., Vigevani, A.: Tetrahedron *22*, 3061 (1966).
50. Casu, B., Reggiani, M., Gallo, G. G., Vigevani, A.: Chem. Soc. (London), Spec. Publ., No. 23, 217 (1968).
51. VanEtten, R. L., Clowes, G. A., Sebastian, J. F., Bender, M. L.: J. Amer. Chem. Soc. *89*, 3253 (1967).
52. Chin, T.-F., Chung, P.-H., Lach, J. L.: J. Pharm. Sci. *57*, 44 (1968).
53. Van Hooidonk, C., Groos, C. C.: Rec. Trav. Chim. Pays-Bas *89*, 845 (1970).
54. Manor, P. C., Saenger, W.: Nature (London) *237*, 392 (1972).
55. Manor, P. C., Saenger, W.: J. Amer. Chem. Soc. *96*, 3630 (1974).
56. Saenger, W., Noltemeyer, M., Manor, P. C., Hingerty, B., Klar, B.: Bioorg. Chem. *5*, 187 (1976).
57. Rees, D. A.: J. Chem. Soc. *B*, 877 (1970).
58. French, D.: Adv. Carbohyd. Chem. *12*, 189 (1957).
59. Takahashi, K., Ono, S.: J. Biochem. (Tokyo) *72*, 679 (1972).
60. Szejtli, J., Budai, Zs.: Acta Chim. Acad. Sci. Hung. *91*, 73 (1976).
61. Swanson, M. A., Cori, C. F.: J. Biol. Chem. *172*, 797 (1948).
62. Myrbäck, K., Järneström, T.: Arkiv Kemi. *1*, 129 (1949).
63. Suetsugu, N., Koyama, S., Takeo, K., Kuge, T.: J. Biochem. (Tokyo) *76*, 57 (1974).
64. Miyazaki, Y.: Saga Daigaku Nogaku Iho *19*, 27 (1972).
65. Cramer, F.: Revs. Pure Appl. Chem. *5*, 143 (1955).
66. Cramer, F.: Angew. Chem. *68*, 115 (1956).
67. Lüttringhaus, A., Cramer, F., Prinzbach, H., Henglein, F. M.: Ann. Chem. *613*, 185 (1958).
68. Wojcik, J. F., Rohrbach, R. P.: J. Phys. Chem. *79*, 2251 (1975).
69. Cramer, F., Henglein, F. M.: Angew. Chem. *68*, 649 (1956).
70. Siegel, B., Breslow, R.: J. Amer. Chem. Soc. *97*, 6869 (1975).
71. Lammers, J. N. J. J., Van Diemen, A. J. G.: Recl. Trav. Chim. Pays-Bas *91*, 733 (1972).
72. Lammers, J. N. J. J.: Recl. Trav. Chim. Pays-Bas *91*, 1163 (1972).
73. Schlenk, H., Sand, D. M.: J. Amer. Chem. Soc. *83*, 2312 (1961).
74. VanEtten, R. L., Sebastian, J. F., Clowes, G. A., Bender, M. L.: J. Amer. Chem. Soc. *89*, 3242 (1967).
75. Koizumi, K., Matsui, K., Higuchi, K.: Yakugaku Zasshi *94*, 1515 (1974).
76. Ogata, N., Sanui, K., Wada, J.: J. Polym. Sci., Polym. Lett. Ed. *14*, 459 (1976).
77. Cramer, F., Henglein, F. M.: Chem. Ber. *90*, 2561 (1957).
78. Demarco, P. V., Thakkar, A. L.: Chem. Commun. *1970*, 2.
79. Thakkar, A. L., Demarco, P. V.: J. Pharm. Sci. *60*, 652 (1971).
80. Wood, D. J., Hruska, F. E., Saenger, W.: J. Amer. Chem. Soc. *99*, 1735 (1977).
81. Otagiri, M., Uekama, K., Ikeda, K.: Chem. Pharm. Bull. *23*, 188 (1975).
82. Bergeron, R., Rowan, R. III.: Bioorg. Chem. *5*, 425 (1976).
83. Bergeron, R., Channing, M. A.: Bioorg. Chem. *5*, 437 (1976).
84. a) Bergeron, R., Channing, M. A., Gibeily, G. J., Pillor, D. M.: J. Amer. Chem. Soc. *99*, 5146 (1977); b) Bergeron, R., Channing, M. A., McGovern, K. A.: private communication. c) Bergeron, R., McPhie, P.: Bioorg. Chem. *6*, in press (1977).

85. Thoma, J. A., French, D.: J. Amer. Chem. Soc. *80*, 6142 (1958).
86. Ikeda, K., Uekama, K., Otagiri, M.: Chem. Pharm. Bull. *23*, 201 (1975).
87. Cramer, F., Saenger, W., Spatz, H.-Ch.: J. Amer. Chem. Soc. *89*, 14 (1967).
88. Broser, W., Lautsch, W.: Z. Naturforsch *8B*, 711 (1953).
89. Casu, B., Rava, L.: Ric. Sci. *36*, 733 (1966).
90. Hoffman, J. L., Bock, R. M.: Biochemistry *9*, 3542 (1970).
91. Seliskar, C. J., Brand, L.: Science *171*, 799 (1971).
92. Kondo, H., Nakatani, H., Hiromi, K.: J. Biochem. (Tokyo) *79*, 393 (1976).
93. Kinoshita, T., Iinuma, F., Tsuji, A.: Biochem. Biophys. Res. Commun. *51*, 666 (1973).
94. Kinoshita, T., Iinuma, F., Tsuji, A.: Chem. Pharm. Bull. *22*, 2413 (1974).
95. Kinoshita, T., Iinuma, F., Atsumi, K., Kanada, Y., Tsuji, A.: Chem. Pharm. Bull. *23*, 1166 (1975).
96. Kinoshita, T., Iinuma, F., Tsuji, A.: Chem. Pharm. Bull. *22*, 2735 (1974).
97. Farmoso, C.: Biochem. Biophys. Res. Commun. *50*, 999 (1973).
98. Farmoso, C.: Biopolymers *13*, 909 (1974).
99. Sensse, K., Cramer, F.: Chem. Ber. *102*, 509 (1969).
100. Takeo, K., Kuge, T.: Stärke *24*, 281 (1972).
101. Harata, K., Uedaira, H.: Bull. Chem. Soc. Japan. *48*, 375 (1975).
102. Thakkar, A. L., Kuehn, P. B., Perrin, J. H., Wilham, W. L.: J. Pharm. Sci. *61*, 1841 (1972).
103. Otagiri, M., Ikeda, K., Uekama, K., Ito, O., Hatano, M.: Chem. Lett. *679* (1974).
104. Uekama, K., Otagiri, M., Kanie, Y., Tanaka, S., Ikeda, K.: Chem. Pharm. Bull. *23*, 1421 (1975).
105. Harata, K., Uedaira, H.: Sen'i Kobunshi Zairyo Kenkyusho Kenkyu Hokoku *112*, 1 (1976).
106. Connors, K. A., Lipari, J. M.: J. Pharm. Sci. *65*, 379 (1976).
107. Paton, R. M., Kaiser, E. T.: J. Amer. Chem. Soc. *92*, 4723 (1970).
108. Flohr, K., Paton, R. M., Kaiser, E. T.: Chem. Commun. *1971*, 1621.
109. Atherton, N. M., Strach, S. J.: J. Chem. Soc., Faraday Trans. 1, *71*, 357 (1975).
110. Flohr, K., Paton, R. M., Kaiser, E. T.: J. Amer. Chem. Soc. *97*, 1209 (1975).
111. Atherton, N. M., Strach, S. J.: J. Magn. Reson. *17*, 134 (1975).
112. Yamaguchi, S., Miyagi, C., Yamakawa, Y., Tsukamoto, T.: Nippon Kagaku Kaishi *1975*, 562.
113. Yamaguchi, S., Tsukamoto, T.: Nippon Kagaku Kaishi *1976*, 1856.
114. Rohrbach, R. P., Rodriguez, L. J., Eyring, E. M., Wojcik, J. F.: J. Phys. Chem. *81*, 944 (1977).
115. Behr, J. P., Lehn, J. M.: J. Phys. Colloq. (Paris) *1973*, 55.
116. Behr, J. P., Lehn, J. M.: J. Amer. Chem. Soc. *98*, 1743 (1976).
117. V. Dietrich, H., Cramer, F.: Chem. Ber. *87*, 806 (1954).
118. McMullan, R. K., Saenger, W., Fayos, J., Mootz, D.: Carbohyd. Res. *31*, 37 (1973).
119. McMullan, R. K., Saenger, W., Fayos, J., Mootz, D.: Carbohyd. Res. *31*, 211 (1973).
120. Saenger, W., Noltemeyer, M.: Angew. Chem. Int. Ed. Engl. *13*, 552 (1974).
121. Saenger, W., Beyer, K., Manor, P. C.: Acta. Cryst. *B32*, 120 (1976).
122. Saenger, W., McMullan, R. K., Fayos, J., Mootz, D.: Acta Cryst. *B, 30*, 2019 (1974).
123. Saenger, W., Noltemeyer, M.: Chem. Ber. *101*, 503 (1976).
124. Noltemeyer, M., Saenger, W.: Nature (London) *259*, 629 (1976).
125. Hingerty, B., Saenger, W.: J. Amer. Chem. Soc. *98*, 3357 (1976).
126. Harata, K.: Bull. Chem. Soc. Japan. *49*, 1493 (1976).
127. Harata, K.: Bull. Chem. Soc. Japan. *49*, 2066 (1976).
128. Cramer, F., Windel, H.: Chem. Ber. *89*, 354 (1956).
129. Cramer, F., Bergmann, U., Manor, P. C., Noltemeyer, M., Saenger, W.: Justus Liebigs Ann. Chem. *1976*, 1169.
130. Harata, K., Uedaira, H.: Nature (London) *253*, 190 (1975).
131. Harata, K.: Bull. Chem. Soc. Japan. *48*, 2409 (1975).
132. Saenger, W.: Jerusalem Symp. Quantum Chem. Biochem. VIII. 265 (1976).
133. Rao, V. S. R., Sundararajan, P. R., Ramakrishnan, C., Ramachandran, G. N.: "Conformation of Biopolymers", Vol II. New York: Academic Press 1967.
134. Giacomini, M., Pullman, B., Maigret, B.: Theoret. Chim. Acta (Berlin) *19*, 347 (1970).

135. Matsui, Y., Kurita, T., Yagi, M., Okayama, T., Mochida, K., Date, Y.: Bull. Chem. Soc. Japan *48*, 2187 (1975).

136. Mochida, K., Matsui, Y.: Chem. Lett. *1976*, 963.

137. Ohnishi, M.: J. Biochem. (Tokyo) *69*, 181 (1971).

138. Mora, S., Simon, I., Elodi, P.: Mol. Cell Biochem. *4*, 205 (1974).

139. Simon, I., Mora, S., Elodi, P.: Mol. Cell Biochem. *4*, 211 (1974).

140. Marshall, J. J.: FEBS Lett. *37*, 269 (1973).

141. a) Hoffman, J. L., J. Macromol. Sci., Chem. *7*, 1147 (1973); b) Cornelius, R. D., Cleland, W. W.: Absts. of ACS meeting, Biolog. Chem. *13*, Sept. 1977.

142. Wiedenhof, N.: Stärke *21*, 163 (1969).

143. Wiedenhof, N., Trieling, R. G.: Stärke *23*, 129 (1971).

144. Vretblad, P.: FEBS Lett. *47*, 86 (1974).

145. Thoma, J. A., French, D.: J. Phys. Chem. *62*, 1603 (1958).

146. Benesi, H. A., Hildebrand, J. H.: J. Amer. Chem. Soc. *71*, 2703 (1949).

147. Turner, D. C., Brand, L.: Biochemistry *7*, 3381 (1968).

148. a) Tabushi, I., Shimokawa, K., Shimizu, N., Shirakata, H., Fujita, K.: J. Amer. Chem. Soc. *98*, 7855 (1976); b) Tabushi, I., Fujita, K., Yuan, L. C.: Tetrahedron Let. 2503 (1977); c) Tabushi, I., Kiyosuke, Y., Sugimoto, T., Yamamura, K.: J. Amer. Chem. Soc., in press.

149. Cohen, J., Lach, J. L.: J. Pharm. Sci. *52*, 132 (1963).

150. Lach, J. L., Cohen, J.: J. Pharm. Sci. *52*, 137 (1963).

151. Lach, J. L., Chin, T.-F.: J. Pharm. Sci. *53*, 69 (1964).

152. Lach, J. L., Chin, T.-F.: J. Pharm. Sci. *53*, 924 (1964).

153. Pauli, W. A., Lach, J. L.: J. Pharm. Sci. *54*, 1745 (1965).

154. Lach, J. L., Pauli, W. A.: J. Pharm. Sci. *55*, 32 (1966).

155. Takeo, K., Kuge, T.: Stärke *24*, 331 (1972).

156. Lewis, E. A., Hansen, L. D.: J. Chem. Soc., Perkin Trans. II. 2081 (1973).

157. Otagiri, M., Miyaji, T., Uekama, K., Ikeda, K.: Chem. Pharm. Bull. *24*, 1146 (1976).

158. Straub, T. S., Bender, M. L.: J. Amer. Chem. Soc. *94*, 8881 (1972).

159. van Hooidonk, C., Breebaart-Hansen, J. C. A. E.: Rec. Trav. Chim. Pays-Bas *90*, 680 (1971).

160. Richards, F. M.: Ann. Rev. Biochem. *32*, 269 (1963).

161. Kauzmann, W.: Adv. Protein Chem. *14*, 1 (1959).

162. Wishnia, A., Pinder, T. W. Jr.: Biochemistry *5*, 1534 (1966).

163. Cramer, F., Dietsche, W.: Chem. Ber. *92*, 1739 (1959).

164. Cramer, F., Dietsche, W.: Chem. and Ind. (London) *1958*, 892.

165. Cramer, F.: Rec. Trav. Chim. Pays-Bas *75*, 891 (1956).

166. Buckingham, A. D.: Quart. Rev. Chem. Soc. *13*, 183 (1959).

167. London, F.: Z. Phys. *63*, 245 (1930).

168. Lippert., J. L., Hanna, M. W., Trotter, P. J.: J. Amer. Chem. Soc. *91*, 4035 (1969).

169. Komiyama, M., Hirai, H.: J. Polym. Sci., Polym. Chem. Ed. *14*, 2009 (1976).

170. Matsui, Y., Naruse, H., Mochida, K., Date, Y.: Bull. Chem. Soc., Japan *43*, 1909 (1970).

171. van Hooidonk, C., Breebaart-Hansen, J. C. A. E.: Recl. Trav. Chim. Pay-Bas *91*, 958 (1972).

172. Wishnia, A., Lappi, S. J.: J. Mol. Biol. *82*, 77 (1974).

173. Bergeron, R. J., Meeley, M. P.: Bioorg. Chem. *5*, 197 (1976).

174. Emert, J., Breslow, R.: J. Amer. Chem. Soc. *97*, 670 (1975).

175. Bender, M. L., VanEtten, R. L., Clowes, G. A., Sebastian, J. F.: J. Amer. Chem. Soc. *88*, 2318 (1966).

176. Bender, M. L., VanEtten, R. L., Clowes, G. A.: J. Amer. Chem. Soc. *88*, 2319 (1966).

177. Bender, M. L., Trans. N. Y.: Acad. Sci. *29*, 301 (1967).

178. Tutt, D. E., Schwartz, M. A.: Chem. Commun. *1970*, 113.

179. Tutt, D. E., Schwartz, M. A.: J. Amer. Chem. Soc. *93*, 767 (1971).

180. Komiyama, M., Bender, M. L.: Bioorg. Chem. *6*, 323 (1977).

181. Komiyama, M., Bender, M. L.: J. Amer. Chem. Soc. *99*, 8021 (1977).

182. Cramer, F.: Angew. Chem. *73*, 49 (1961).

183. Hennrich, N., Cramer, F.: Chem. and Ind. (London) *1961*, 1224.

184. Hennrich, N., Cramer, F.: J. Amer. Chem. Soc. *87*, 1121 (1965).

185. Brass, H. J., Bender, M. L.: J. Amer. Chem. Soc. *95*, 5391 (1973).
186. van Hooidonk, C., Breebaart-Hansen, J. C. A. E.: Rec. Trav. Chim. Pays-Bas *89*, 289 (1970).
187. Congdon, W. I., Bender, M. L.: Bioorg. Chem. *1*, 424 (1971).
188. Griffiths, D. W., Bender, M. L.: J. Amer. Chem. Soc. *95*, 1679 (1973).
189. Cramer, F., Kampe, W.: Tetrahedron Let. 353 (1962).
190. Cramer, F., Kampe, W.: J. Amer. Chem. Soc. *87*, 1115 (1965).
191. Straub, T. S., Bender, M. L.: J. Amer. Chem. Soc. *94*, 8875 (1972).
192. Cramer, F.: Chem. Ber. *86*, 1576 (1953).
193. Breslow, R., Campbell, P.: J. Amer. Chem. Soc. *91*, 3085 (1969).
194. Breslow, R., Campbell, P.: Bioorg. Chem. *1*, 140 (1971).
195. Breslow, R., Kohn, H., Siegel, B.: Tetrahedron Let. 1645 (1976).
196. a) Szejtli, J., Banky-Elod, E.: Acta Chim. Acad. Sci. Hung. *91*, 67 (1976); b) idem., Stärke *27*, 368 (1975).
197. Tabushi, I., Fujita, K., Kawakubo, H.: J. Amer. Chem. Soc. *99*, 6456 (1977).
198. Cramer, F.: Chem. Ber. *84*, 851 (1951); Ann. Chem. *579*, 17 (1953).
199. Schlenk, H., Sand, D. M., Tillotson, J. A.: J. Amer. Chem. Soc. *77*, 3587 (1955).
200. Matsui, Y., Naruse, H., Mochida, K., Date, Y.: Bull. Chem. Soc. Japan. *43*, 1910 (1970).
201. Van der Jagt, D. L., Killian, F. L., Bender, M. L.: J. Amer. Chem. Soc. *92*, 1016 (1970).
202. Matsui, Y., Mochida, K., Fukumoto, O., Date, Y.: Bull. Chem. Soc. Japan. *48*, 3645 (1975).
203. Yamada, K., Morimoto, J., Iida, H.: Chiba Daigaku Kogakubu Kenkyu Hokoku *24*, 71 (1973).
204. Yamada, K., Morimoto, J., Iida, H.: Chiba Daigaku Kogakubu Kenkyu Hokoku *26*, 85 (1975).
205. Sunamoto, J., Okamoto, H., Taira, K., Murakami, Y.: Chem. Lett. *1975*, 371.
206. Hamada, Y., Nambu, N., Nagai, T.: Chem. Pharm. Bull. *23*, 1205 (1975).
207. Kurozumi, M., Nambu, N., Nagai, T.: Chem. Pharm. Bull. *23*, 3062 (1975).
208. Koizumi, K., Fujimura, K.: Yakugaku Zasshi *92*, 32 (1972).
209. Shibusawa, T., Hamayose, T., Sasaki, M.: Nippon Kagaku Kaishi. *12*, 2171 (1975).
210. Trommsdorff, H., G. m. b. H., Co. K.-G. Arzneimittelfabrik: Fr. Demande 2,209,582; Chem. Abstr. *82*, 144957h (1975).
211. Suzuki, Y., Ikura, H.; Japan Kokai Patent 75 58,208: Chem. Abstr. *83*, 183396q (1975).
212. Mifune, A., Katsuda, Y., Yoneda, T.; Ger. Offen. 2,357,826: Chem. Abstr. *82*, 39586p (1975).
213. Kawamura, S., Murakami, M., Kawada, H.; Japan Kokai Patent 75 100,217: Chem. Abstr. *83*, 183412s (1975).
214. Ikeda, K., Uekama, K.; Japan Kokai Patent 76 48,420: Chem. Abstr. *85*, 51767t (1976).
215. Katsuda, Y., Yamamoto, S.; Japan Kokai Patent 76 81,888: Chem. Abstr. *85*, 138639b (1976).
216. Nagai, T.; Japan Kokai Patent 75 116,617: Chem. Abstr. *84*, 111654v (1976).
217. Hayashi, M., Ishihara, A.; Brit. Patent 1,419,221: Chem. Abstr. *84*, 121295t (1976).
218. Hamuro, J., Akiyama, M.; Japan Kokai Patent 75 36,422: Chem. Abstr. *83*, 79544a (1975).
219. Shimada, K., Tanaka, I., Fukada, T., Nagahama, S.; Japan Kokai Patent 75 40,726: Chem. Abstr. *83*, 73533z (1975).
220. Suetani, T., Inaba, K.; Japan Kokai Patent 75 35,319: Chem. Abstr. *83*, 65479w (1975).
221. Hirano, S., Tsumura, J., Imazaki, I., Ohuchi, M., Kito, H.; Japan Kokai Patent 75 64,418: Chem. Abstr. *83*, 152403k (1975).
222. Hirano, S., Tsumura, J., Izeki, I., Kawamura, H., Ohhara, M.; Japan Kokai Patent 75 63,126: Chem. Abstr. *83*, 120672f (1975).
223. Suzuki, Y., Ikura, H.; Japan Kokai Patent 75 83,454: Chem. Abstr. *83*, 146032x (1975).
224. Kawamura, S., Murakami, M., Kawata, H., Terao, J.; Japan Kokai Patent 75 94,108: Chem. Abstr. *83*, 197823p (1975).
225. Hayashi, M., Ishihara, A.; S. African Patent 74 00,295: Chem. Abstr. *83*, 178420p (1975).
226. Takamura, S., Murakami, M., Kawata, H., Kawamura, S.; Japan Kokai Patent 75 89,516: Chem. Abstr. *83*, 152384e (1975).

227. Shimada, K., Tanaka, I., Fukada, T.; Japan Kokai Patent 75 46,826 : Chem. Abstr. *83*, 92413b (1975).
228. Hashimoto, S., Katayama, A.; Japan Kokai Patent 75 35,349: Chem. Abstr. *83*, 7607a (1975).
229. Noda, K., Furuya, K., Miyata, S., Tosu, S., Yoneda, T.; Ger. Offen. 2,356,098: Chem. Abstr. *81*, 54445q (1974).
230. Ogura, T., Saito, Y., Kobayashi, T.; Japan Kokai Patent 75 142,790: Chem. Abstr. *84*, 117800m (1976).
231. Kinoshita, T., Iinuma, F., Tsuji, A.: Anal. Biochem. *61*, 632 (1974).
232. Kinoshita, T., Iinuma, F., Tsuji, A.: Chem. Pharm. Bull. *22*, 2421 (1974).
233. Kinoshita, T.: Kagaku to Seibutsu *13*, 392 (1975).
234. MacNicol, D. D.: Tetrahedron Let. *1975*, 3325.
235. Kurono, Y., Stamoudis, V., Bender, M. L.: Bioorg. Chem. *5*, 393 (1976).
236. Lineweaver, H., Burk, D.: J. Amer. Chem. Soc. *56*, 658 (1934).
237. Eadie, G. S.: J. Biol. Chem. *146*, 85 (1942).
238. Lautsch, W., Wiechert, R., Lehmann, H.: Kolloid-Z. *135*, 134 (1954).
239. Staerk, J., Schlenk, H. Jr.: Abstracts, 149th National Meeting of the American Chemical Society, Detroit, Mich., p11c, 1965.
240. Komiyama, M., Breaux, E. J., Bender, M. L.: Bioorg. Chem. *6*, 127 (1977).
241. Bell, R. P., Kuhn, A. T.: Trans. Faraday Soc. *59*, 1789 (1963).
242. a) Bender, M. L.: "Mechanisms of Homogeneous Catalysis from Protons to Proteins". Wiley-Interscience, New York, N. Y. (1971); b) Komiyama, M., Bender, M. L.: J. Amer. Chem. Soc., Submitted for publication (1977); c) Komiyama, M., Bender, M. L.: J. Amer. Chem. Soc., Submitted for publication (1977).
243. Bender, M. L., Pollock, E. J., Neveu, M. C.: J. Amer. Chem. Soc. *84*, 595 (1962).
244. Bender, M. L., Turnquest, B. W.: J. Amer. Chem. Soc. *79*, 1652 and 1656 (1957).
245. Bruice, T. C., Schmir, G. L.: J. Amer. Chem. Soc. *79*, 1663 (1957).
246. Stadtman, E. R.: "The Mechanism of Enzyme Action". McElroy, W. D., Glass, B. (ed.) p. 581. Baltimore: Johns Hopkins Press 1954.
247. Jencks, W. P., Carriuolo, J.: J. Biol. Chem. *234*, 1272 and 1280 (1959).
248. Mochida, K., Matsui, Y., Ota, Y., Arakawa, K., Date, Y.: Bull. Chem. Soc. Japan *49*, 3119 (1976).
249. Brown, B. R.: Quart. Rev. Chem. Soc. *5*, 131 (1951).
250. Hall, G. A. Jr., Verhoek, F. H.: J. Amer. Chem. Soc. *69*, 613 (1947).
251. Kemp, D. S., Paul, K.: J. Amer. Chem. Soc. *92*, 2553 (1970).
252. Thomson, A.: J. Chem. Soc. *B*, 1198 (1970).
253. Bardsley, J., Baugh, P. J., Goodall, J. I., Phillips, G. O.: Chem. Commun. *1974*, 890.
254. Griffiths, D. W., Bender, M. L.: Bioorg. Chem. *4*, 84 (1975).
255. Bruice, T. C., Brandbury, W. C.: J. Amer. Chem. Soc. *87*, 4846 (1965).
256. Bruice, T. C., Turner, A.: J. Amer. Chem. Soc. *92*, 3422 (1970).
257. Milstien, S., Cohen, L. A.: Proc. Nat. Acad. Sci. USA *67*, 1143 (1970).
258. Page, M. I., Jencks, W. P.: Proc. Nat. Acad. Sci. USA *68*, 1678 (1971).
259. Olah, G. A., Dunne, K., Kelley, D. P., Mo, Y. K.: J. Amer. Chem. Soc. *94*, 7438 (1972).
260. Mitani, M., Tsuchida, T., Koyama, K.: Chem. Commun. *1974*, 869.
261. Ohara, M., Watanabe, K.: Angew. Chem. Intern. Ed. *14*, 820 (1975).
262. Machida, Y., Bergeron, R., Flick, P., Bloch, K.: J. Biol. Chem. *248*, 6246 (1973).
263. Bergeron, R., Machida, Y., Bloch, K.: J. Biol. Chem. *250*, 1223 (1975).
264. Gray, G. R., Ballou, C. E.: J. Biol. Chem. *246*, 6835 (1971).
265. Ilton, M., Jevans, A. W., McCarthy, E. D., Vance, D., White, H. B. III, Bloch, K.: Proc. Nat. Acad. Sci, USA *68*, 87 (1971).
266. van Hooidonk, C.: Rec. Trav. Chim. Pay-Bas *91*, 1103 (1972).
267. Cramer, F., Dietsche, W.: Chem. Ber. *92*, 378 (1959).
268. Mikolajczyk, M., Drabowicz, J., Cramer, F.: Chem. Commun. *1971*, 317.
269. Benschop, H. P., Van den Berg, G. R.: Chem. Commun. *1970*, 1431.
270. Mikolajczyk, M., Drabowicz, J.: Tetrahedron Let. *1972*, 2379.

271. Kitaura, Y., Bender, M. L.: Bioorg. Chem. *4*, 237 (1975).
272. Woodward, R. B., Cava, M. P., Ollis, W. D., Hunger, A., Daeniker, H. U., Schenker, K.: Tetrahedron *19*, 247 (1963).
273. Komiyama, M., Bender, M. L.: Proc. Nat. Acad. Sci. USA *73*, 2969 (1976).
274. Kitano, H., Okubo, T.: J. Chem. Soc., Perkin II *1977*, 432.
275. Koshland, D. E. Jr., Strumeyer, D. H., Ray, W. J. Jr.: Brookhaven Symp. Biol. *15*, 101 (1962).
276. Weil, L., Buchert, A. R.: Federation Proc. *11*, 307 (1952).
277. Schoellmann, G., Shaw, E.: Biochemistry *2*, 252 (1963).
278. Bender, M. L., Clement, G. E., Kezdy, F. J., Heck, H. D'A.: J. Amer. Chem. Soc. *86*, 3680 (1964).
279. Cramer, F., Mackensen, G.: Angew. Chem. *78*, 641 (1966).
280. Cramer, F., Mackensen, G.: Chem. Ber. *103*, 2138 (1970).
281. Iwakura, Y., Uno, K., Toda, F., Onozuka, S., Hattori, K., Bender, M. L.: J. Amer. Chem. Soc. *97*, 4432 (1975).
282. Blow, D. M., Birktoft, J. J., Hartley, B. S.: Nature (London) *221*, 337 (1969).
283. Wright, C. S., Alden, R. A., Kraut, J.: Nature (London) *221*, 235 (1969).
284. Komiyama, M., Bender, M. L.: Bioorg. Chem. *6*, 13 (1977).
285. Gruhn, W. B., Bender, M. L.: Bioorg. Chem. *3*, 324 (1974).
286. Gruhn, W. B., Bender, M. L.: J. Amer. Chem. Soc. *91*, 5883 (1969).
287. Gruhn, W. B., Bender, M. L.: Bioorg. Chem. *4*, 219 (1975).
288. van Hooidonk, C., de Korte, D. C., Reuland-Meereboer, M. A. C.: Rec. Trav. Chim. Pay-Bas *96*, 25 (1977).
289. Siegel, B., Pinter, A., Breslow, R.: J. Amer. Chem. Soc. *99*, 2309 (1977).
290. Lammers, J. N. J. J., Koole, J. L., Hurkmans, J.: Stärke *23*, 167 (1971).
291. Takeo, K., Yagi, F., Kuge, T.: Kyoto Furitsu Daigaku Gakujutsu Hokoku, Nogaku *1972*, 159.
292. Bines, B. J., Whelan, W. J.: Chem. and Ind. 997 (1960).
293. Umezawa, S., Tatsuta, K.: Bull. Chem. Soc. Japan *41*, 464 (1968).
294. Melton, L. D., Slessor, K. N.: Carbohyd. Res. *18*, 29 (1971).
295. Takeo, K., Sumimoto, T., Kuge, T.: Stärke *26*, 111 (1974).
296. Koester, R., Amen K. L., Dahlhoff, W. V.: Justus Liebigs Ann. Chem. 752 (1975).
297. Kurita, H., Kawazu, M., Takashima, K.; Japan Kokai Patent 74 85,014: Chem. Abstr. *81*, 169767p (1974).
298. Takeo, K., Kuge, T.: Stärke *28*, 226 (1976).
299. a) Tabushi, I., Shimizu, N., Sugimoto, T., Shiozuka, M., Yamamura, K.: J. Amer. Chem. Soc. *99*, 7100 (1977); b) Tabushi, I., Shimokawa, K., Fujita, K.: Tetrahedron Let. 1527 (1977).
300. Frue, M., Harada, A., Nozakura, S.: J. Polym. Sci., Polym. Let. Ed. *13*, 357 (1975).
301. Harada, A., Furue, M., Nozakura, S.: Macromolecules *9*, 701 (1976).
302. Harada, A., Furue, M., Nozakura, S.: Macromolecules *9*, 705 (1976).
303. Harada, A., Furue, M., Nozakura, S.: Macromolecules *10*, 676 (1977).
304. Breslow, R., Fairweather, R., Keana, J. F.: J. Amer. Chem. Soc. *89*, 2135 (1967).
305. Breslow, R., Overman, L. E.: J. Amer. Chem. Soc. *92*, 1075 (1970).
306. Breslow, R., Chipman, D.: J. Amer. Chem. Soc. *87*, 4195 (1965).
307. Matsui, Y., Yokoi, T., Mochida, K.: Chem. Let. *1976*, 1037.

Author Index

Author Index

Subject Index

Topics
in
Current Chemistry

Fortschritte der chemischen Forschung

Managing Editor: F. L. Boschke

Springer-Verlag
Berlin
Heidelberg
New York

Reactivity and Structure

Concepts in Organic Chemistry

Editors: K. Hafner, J.-M. Lehn, C. W. Rees,
P. v. Ragué Schleyer, B. M. Trost, R. Zahradník

Volume 1: J. Tsuji

Organic Synthesis by Means of Transition Metal Complexes

A Systematic Approach

4 tables. IX, 199 pages. 1975
ISBN 3-540-07227-6

Contens: Comparison of synthetic reactions by transition metal complexes with those by Grignard reagents. – Formation of σ-bond involving transition metals. – Reactivities of σ-bonds involving transition metals. – Insertion reactions. – Liberation of organic compounds from the σ-bonded complexes. – Cyclization reactions, and related reactions. – Concluding remarks.

Volume 2: K. Fukui

Theory of Orientation and Stereoselection

72 figures, 2 tables. VII, 134 pages. 1975
ISBN 3-540-07426-0

Contens: Molecular Orbitals. – Chemical Reactivity Theory. – Interaction of Two Reacting Species. – Principles Governing the Reaction Pathway. – General Orientation Rule. – Reactivity Indices. – Various Examples. – Singlet-Triplet Selectivity. – Pseudoexcitation. – Three-species Interaction. – Orbital Catalysis. – Thermolytic Generation of Excited States. – Reaction Coordinate Formalism. – Correlation Diagram Approach. – The Nature of Chemical Reactions.
Appendix 1: Principles Governing the Reaction Path – An MO-Theoretical Interpretation. – Appendix 2: Orbital Interaction between Two Molecules.

Volume 3: H. Kwart, K. King

d-Orbitals in the Chemistry of Silicon, Phosphorus and Sulfur

Approx. 220 pages. 1977
ISBN 3-540-07953-X
Contents: Theoretical Basis for d-Orbital Involvement. – Physical Properties Related to dp-π Bonding. – The Effects of dp-π Bonding on Chemical Properties and Reactivity. – Pentacovalency.

Volume 4: W. P. Weber, G. W. Gokel

Phase Transfer Catalysis in Organic Synthesis

XV, 280 pages. 1977
ISBN 3-540-08377-4

Contents: Introduction and Principles. – The Reaction of Dichlorocarbene with Olefins. – Reactions of Dichlorocarbene with Non-Olefinic Substrates. – Dibromocarbene and Other Carbenes. – Synthesis of Ethers. – Synthesis of Esters. – Reactions of Cyanide Ion. – Reactions of Superoxide Ions. – Reactions of Other Hucleophiles. – Alkylation Reactions. – Oxidation Reactions. – Reduction Techniques. – Preparation and Reactions of Sulfur Containing Substrates. – Ylids. – Altered Reactivity. Addendum: Recent Developments in Phase Transfer Catalysis.

Volume 5: N. D. Epiotis

Theory of Organic Reactions

Approx. 70 figures. Approx. 340 pages. 1978
ISBN 3-540-08551-3

Contents: One determinental theory of chemical reactivity. – Configuration interaction overview of chemical reactivity. The dynamic linear combination of fragment configurations method. – Even-even intermolecular multicentric reactions. – The problem of correlation imposed barriers. – Reactivity trends of thermal cycloadditions. – Reactivity trends of singlet photochemical cycloadditions. – Miscellaneous intermolecular multicentric reactions. – π+σ addition reactions. – Even-odd multicentric intermolecular reactions. – Potential energy surfaces for odd-odd multicentric intermolecular reactions. – Even-even intermolecular bicentric reactions. – Even-odd intermolecular bicentric reactions. – Odd-odd intermolecular bicentric reactions. Potential energy surfaces for geometric isomerization and radical combination. – Odd-odd intramolecular multicentric reactions. – Even-even intramolecular multicentric reactions. – Mechanisms of electrocyclic reactions. – Triplet reactivity. – Photophysical processes. – The importance of low lying nonvalence orbitals. – Divertissements. – A contrast of "accepted" concepts of organic reactivity and the present work.

Springer-Verlag
Berlin Heidelberg New York